Main Line Passenger Trains
In and Around London

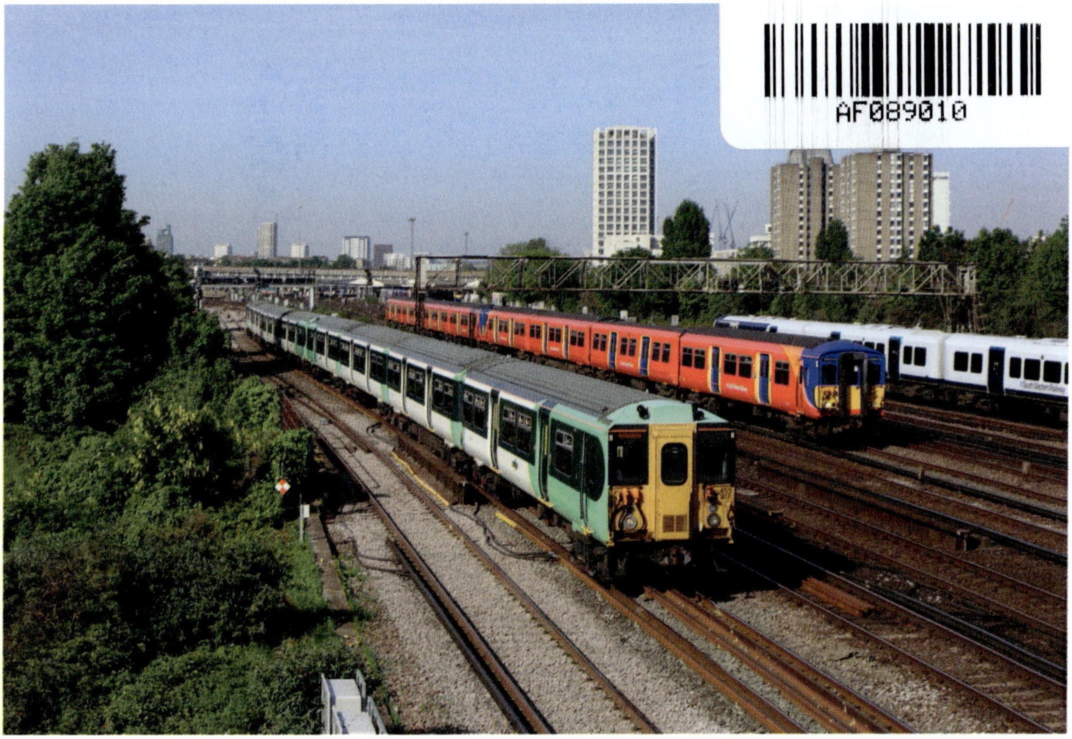

JAMIE SQUIBBS

KEY Books

BRITAIN'S RAILWAYS SERIES

Front cover image: Thameslink 700155 *Trainbow* working the 9J09 06.24 Peterborough to Horsham service, pulling away from London Bridge on 11 July 2022. Also passing are 375911, 375604 and 375926 working the 1G89 06.15 Broadstairs to London Cannon Street service and 465010 and 465190 (behind) working 1D54 06.55 Strood to London Charing Cross.

Back cover image: South Western Railway 450044 leads 450069 and 450027, working the 2F88 07.25 Woking to London Waterloo service on the Nine Elms to Waterloo Viaduct, east of Vauxhall on 26 August 2022.

Title page image: Southern 455817 and 455814 working the 2B23 08.09 Epsom Downs to London Victoria service, running parallel with South Western Railway (SWR) 455750 and 455864 working the 2M18 08.34 Chessington South to London Waterloo service, pulling away from Clapham Junction on 6 May 2022. Both Southern 455s have since been scrapped with a similar fate likely for the SWR 455s, when they are eventually replaced by Class 701s. 455750 was formerly 455743 but was renumbered in 1991 to commemorate Wimbledon depot receiving British Standard 5750 quality accreditation.

Contents page image: In a tribute to HM Queen Elizabeth II (1926–2022), DB Cargo 67007, in special Platinum Jubilee colours, works the 1Y46 08.39 Victoria to Ashford International Belmond British Pullman (67024 on the rear) crossing the River Thames at Grosvenor Bridge on 9 October 2022.

Published by Key Books
An imprint of Key Publishing Ltd
PO Box 100
Stamford
Lincs PE9 1XQ

www.keypublishing.com

The right of Jamie Squibbs to be identified as the author of this book has been asserted in accordance with the Copyright, Designs and Patents Act 1988 Sections 77 and 78.

Copyright © Jamie Squibbs, 2023

ISBN 978 1 80282 267 0

All rights reserved. Reproduction in whole or in part in any form whatsoever or by any means is strictly prohibited without the prior permission of the Publisher.

Typeset by SJmagic DESIGN SERVICES, India.

Contents

Introduction 4

Avanti West Coast 5

C2C 8

Caledonian Sleeper 11

Chiltern Railways 13

East Midlands Railway 17

Elizabeth Line 21

Eurostar 24

Grand Central 26

Great Northern 28

GWR and Heathrow Express 31

Greater Anglia 40

Hull Trains 48

Lumo 50

LNER 51

London Northwestern Railway 56

London Overground 59

Railtours 66

Southeastern 69

Southern and Gatwick Express 75

South Western Railway 83

Thameslink 91

Introduction

It gives me great pride and pleasure to share with you my first book and I would like to thank Key Publishing for giving me this opportunity. I've been interested in railways since I was three years old. I grew up in South Wales and was lucky enough to travel around the UK with my Dad. Using a Chinon CS 35mm SLR, I had a preference for photographing loco-hauled passenger trains, particularly recording the last days of CrossCountry 47s and AC electrics. I have wonderful childhood memories, which include naming 31289 *Phoenix* at the Northampton and Lamport Railway; driving Railtrack 31s at Derby Fragonset; visiting my good friend, Steve Finn, who worked at Pete Waterman's depot in Crewe; visiting Martin Care at Crown Point and seeing the last of my beloved Anglia 86s; driving the EuroShuttle simulator at Cheriton, Kent, and having a cab-ride to Coquelles, near Calais, and back! To improve my photography, I upgraded to Canon DSLR cameras from 2006. I always wanted to write railway books and was inspired by the excellent work of Colin Marsden, who has always been very supportive of my work. After spending three years studying at Loughborough University, I moved to Twickenham in mid-2016 to embark on a new life in London (being nearer to Arsenal was just one of the reasons). At this point, my interest shifted to photographing the entire traction scene, and this is what I aim to showcase. Not only have operators, liveries and traction changed, so has the skyline. Although it is impossible to name everyone who has helped me, I would like to offer my sincere thanks to Philip Sherratt, Colin Marsden, Derek Riley, Jim Dowsett, James Hawkes, Marc Ely, Wayne Walsh, Andrew Lamport, Jack Pallister, Nic Joynson, Nathan Williamson, Fraser Hay, Sam Smith, Neil Thaler, Luca Chalklin, Marcus Langford, Richard Dyke and Amelia Lancaster. I hope you enjoy my photography from around London. Please visit www.jamiesquibbs.co.uk to see more photographs.

Best wishes
Jamie

Avanti West Coast

The Inter City West Coast franchise (ICWC) was operated by Virgin Trains (VT), a joint venture between Stagecoach and Virgin, from its inception on 9 March 1997 until 7 December 2019 when it was superseded by the West Coast Partnership franchise (WCP), with the successful bidder responsible for the running of West Coast Main Line (WCML) services and future HS2 services, at the time of awarding, until at least 2031.

Avanti West Coast – a consortium of FirstGroup (70 percent) and Trenitalia (30 percent) commenced operations on 8 December 2019, ending 22 years of iconic Virgin red liveries on the WCML and inheriting that company's 56 Class 390 Electric Multiple Units (EMUs) and 20 Class 221 Diesel-Electric Multiple Units (DEMU) stock.

The Pendolinos work services along the WCML, which connects Edinburgh and Glasgow in Scotland, with Preston, Carlisle, Blackpool, Manchester and Liverpool in northwest England and Birmingham and Euston (the oldest intercity terminal in London). The 221 Voyagers are used on WCML services to Shrewsbury and North Wales as well as trains from Scotland via Birmingham.

Built by Bombardier in Bruges, 44 Class 221 Voyagers were introduced by Virgin from April 2002, replacing HSTs, Class 47s and MK2 rolling stock. When West Coast CrossCountry services transferred to ICWC in November 2007, VT inherited five-car 221101–118 and four-car 221142–144, with the remaining cars going to Arriva CrossCountry (AXC). In 2010, 221142 and 221143 were converted to five cars using the non-driving vehicles from 221144. The driving vehicles of 221144 later moved to AXC.

VT introduced Class 390s, built by Alstom using Fiat Ferroviaria's Pendolino tilting technology from July 2002, replacing the ageing BR Class 86, 87, 90, MK2 and MK3 rolling stock as part of the 'red revolution'. Originally intended to be just eight cars, 53 nine-car Class 390s (390001–053) were built at Alstom's Washwood Heath depot, northeast of Birmingham city centre, between 2001 and 2004, with an additional four 11-car Pendolinos (390154–157) built at Alstom's Savigliano factory in Italy between 2010 and 2012, one to replace 390033, which was written off following the Grayrigg derailment on 23 February 2007. To further increase capacity, in 2012, 31 of the original nine-car Class 390/0s were extended with two newly built coaches to form 11-car trains and were renumbered into the 390/1 series.

Towards the end of its tenure, Virgin Trains started repainting its Pendolino fleet into a revised 'flowing silk' livery, with 390010 the first to be released in the new livery in September 2017 from the new Alstom factory in Widnes, Cheshire. In April 2019, news broke that Stagecoach (in a consortium with SNCF and Virgin) had been disqualified from bidding for the WCP for not meeting pension rules.

From late 2019, the VT Pendolinos were released without corporate branding, while others were de-branded creating a plain neutral livery. First Trenitalia commenced operation of the West Coast franchise from 8 December 2019. As part of the franchise commitments, Avanti will be carrying out an entire refurbishment of its Pendolino fleet with more seats, reliable wi-fi and improved catering. Avanti will also replace its entire Super Voyager fleet with new Hitachi A-Trains, due to enter service in 2023. These being 13 five-car bi-modal Class 805 and 10 seven-car Class 807 EMUs. 390155 and 156 were the first Pendolinos to receive Avanti West coast colours for the franchise launch with the rest following from mid-2020 onwards.

The UK rail-franchising system was abolished ensuing the Covid-19 pandemic, which saw the WCP replaced with an Emergency Rail Measures Agreement (ERMA) extended until 16 October 2022. It had been expected that First Trenitalia would receive a direct award to operate the West Coast partnership franchise until 2032. However, following poor performance and severely reduced services, Avanti was granted a six-month ERMA until April 2023, during which its performance will be assessed before the Department for Transport (DfT) determines whether to award Avanti a National Rail Contract.

In the original Virgin Trains livery, albeit with grey doors instead of stripes, nine-car 390013 powers the 1H32 15.00 Euston to Manchester Piccadilly service passing the London Underground station of South Kenton in the borough of Brent on 31 May 2017.

For the franchise switch-over, Avanti inherited the Pendolino fleet in a neutral livery, having had the Virgin branding and red silks removed. Awaiting a repaint into Avanti colours, the 11-car 390157 works 1F22 17.03 Euston to Liverpool Lime Street service, passing under the bridge at Northwick Park, which carries the London Underground Metropolitan line and London to Aylesbury line, on 5 June 2020.

Displaying the Avanti West Coast livery, 390050 works 1H12 08.20 Euston to Manchester Piccadilly service passing Bushey Rail Station and crossing the viaduct on 8 July 2022. The eastern side of Bushey viaduct shows a newer bridge alongside the original arches, which is a Grade II listed structure.

Avanti West Coast launched the first fully wrapped Pride train in August 2020, exhibiting its commitment to diversity and inclusion. The train features all 11 stripes of Daniel Quasar's 2018 design of the Progress Pride flag including black, brown, pale blue, pink and white to represent people of colour, transgender people and those living with HIV/AIDS within the LGBTQ+ community. 390119 *Progress* works 5H67 10.19 Wembley Inter City Depot to Euston at Kilburn in readiness to work 9H67 11.19 to Manchester Piccadilly on 16 October 2022.

Virgin Trains 221114 powers 9M52 08.52 Edinburgh to Euston service at South Kenton on 25 February 2019. Only 221101 went into an interim livery prior to the franchise switch over to Avanti. As part of the franchise agreement, new, more environmentally friendly trains would be introduced, as voyagers would spend long distances working on diesel under electric wires.

Between July and December 2020, all 20 Avanti 221s received an extensive interior refresh at Ilford EMUD. Now in interim Avanti West Coast livery, 221116 leads 221112 on 1R13 04.48 Holyhead to Euston at South Kenton on 2 June 2022. The voyagers will soon be replaced by new BMU Class 805 and EMU Class 807 units with 221142 and 221143 the first to be returned to their leasing company, Beacon Rail in June 2022.

C2C

C2C, owned by Trenitalia, operates the Essex Thameside franchise, which incorporates services between Fenchurch Street, London and Shoeburyness, Southend-on-Sea including a loop line via Rainham, Kent and Tilbury, Essex and a branch line via Ockendon, Essex. The franchise was awarded in 1996 to Prism Rail operating as LTS Rail. In 2000, Prism Rail was bought by National Express Group, which renamed its services 'C2C' in 2003 meaning 'City or Capital to Coast' or 'Committed to Customers'. Trenitalia purchased National Express's franchise in February 2017 (the last rail franchise National Express held in the UK) and the current franchise deal runs until 2029.

In addition to Fenchurch Street, C2C uses Liverpool Street at weekends to serve Stratford station to give customers direct access to Westfield Shopping Centre, London Stadium, and easier access to London's West End. C2C can also serve Stratford by diverting trains via the irregularly used Bow Curve without the need to divert into Liverpool Street.

C2C has a core fleet of 74 four-car Class 357 Electrostars, which were built by Adtranz at Derby Litchurch Lane and delivered in two batches, 46 (357001–357046) in 1999–2001 and 28 (357201–228) in 2001–02, replacing the slam-doors Class 310s, Class 312s and enabling the return of Class 317s to West Anglia Great Northern. Therefore, C2C became the first train operating company (TOC) to replace its entire fleet of trains.

With the Leadenhall building in the shape of a cheese grater and 30 St Mary Axe 'Gherkin' building standing out in an array of central London skyscrapers, 375003 leads 375025 working 2B30 12.30 Fenchurch Street to Shoeburyness at Shadwell on 2 April 2022. The DLR from Tower Gateway, just east of Fenchurch Street, runs parallel with the C2C lines until Limehouse where they diverge.

From July 2015, 17 Class 357/2s (357212–357228) received an internal refresh with wider aisles and seating arrangement altering from 3+2 to 2+2, to provide more standing capacity for suburban peak services into Fenchurch Street. These were renumbered into the 357/3 series becoming 357312–328.

In April 2016, C2C signed a deal with the Department for Transport (DfT) for the leasing of six new Bombardier four-car Class 387s (387301–387306) from Porterbrook to provide 13,000 extra seats in peak hours to match growing passenger demand on its network. The lease was initially for three years as a short-term measure before the arrival of newly built trains as promised under its franchise award in 2014. The 387s first entered service in November 2016 and continued until June 2022, although by this time, 387301, 302 and 306 were on loan at Great Western Railway (GWR).

Financed by Porterbrook, C2C announced in December 2017 that it had agreed terms with Bombardier (later bought out by Alstom in January 2021) and the DfT for the procurement of six newly built ten-car *Aventra* EMUs providing a capacity increase of 20 percent and as a direct replacement for the interim 387s. The order was later amended to 12 five-car Class 720s (720601–612). Initially scheduled to enter service by 2021, the Aventra programme across all operators was delayed. The new order is now expected to enter service in 2023.

Passing the Revolution Karting track and Mile End Leisure Centre and Stadium, east of Limehouse, 375019 leads 357316 (ex-357216) working 5E80 09.17 Fenchurch Street to East Ham EMUD, having previously worked 2N09 08.34 peak buster from Laindon, Essex on 15 June 2022. Fenchurch Street has limited capacity with only four platforms, therefore there are several empty coaching stock (ECS) movements to Shoeburyness CSD and East Ham EMUD during the morning peak.

The electrified and now single-lined Bow Curve connects the London, Tilbury and Southend lines at Gas Factory Junction with the Great Eastern Main Line (GEML) at Bow Junction and is used infrequently for diversions in emergency situations or engineering works. After departing Stratford on diversion, 357042 and 357003 work 9B25 14.26 Shoeburyness to Fenchurch Street adjacent to Arnold Road, Bow on 12 June 2022. The location at the rear of the train would have been the site of the former Bow Road station, which closed in November 1949 following the withdrawal of passenger services.

Passing over the District, Hammersmith and City underground lines, 357004 leads 357011 away from Gas Factory Junction and onto the Bow Curve working the diverted 9B36 18.38 Fenchurch Street to Shoeburyness on 12 June 2022. Gas Factory Junction and the LTS lines can be seen to the far right and the skyscrapers of Canary Wharf dominate the background.

Left: The 387s worked peak-time services although their use during the Covid-19 pandemic was limited following a downturn in passenger numbers. On 15 June 2022, 387304 and 387303 having worked 1P13 07.02 Shoeburyness to Fenchurch Street peak-buster, return with 5S30 08.08 to Shoeburyness CSD ECS at Limehouse. These Electrostars would transfer to Great Northern soon after and be replaced in due course by new Class 720s.

Below: Commissioning runs for the new C2C Aventras started in April 2022. Recently released from Derby Litchurch Lane, 720602 and 720604 work 5Q92 13.00 Southend Central to Fenchurch Street test run through Shadwell on 11 July 2022.

Caledonian Sleeper

As part of Scottish franchise relet, the Caledonian sleeper became a stand-alone franchise from 31 March 2015 and was awarded to Serco on an initial 15-year contract. GBRf provided Serco with the locos and drivers for the Anglo-Scottish sleeper. However, GBRf's dedicated Class 92 fleet were struggling with reliability at the start of the franchise, which resulted in Serco continuing with hired-in DB 90s before switching to Freightliner (FL) 90s. GBRf also used ACLG's 86101, 86401 and 87002 for ECS duties.

Caledonian Sleeper (CS) operates two trains in each direction between London Euston and Scotland, Sundays to Fridays. One train is the lowland sleeper to Glasgow Central with a portion for Edinburgh splitting at Carstairs, South Lanarkshire. The other train is the Highland sleeper to Inverness with portions for Fort William and Aberdeen, all dividing at Edinburgh Waverley. Both sleepers start and finish at Euston as a 16-carriage train making them the longest domestic passengers trains in the UK.

Under the new franchise, Serco ordered 75 new state-of-the-art MK5 carriages, built at CAF Beasain in northern Spain, funded by Capital Grants from both the UK and Scottish governments, to replace the BR-era MK2s and MK3s. The new MK5s were introduced on the lowland sleepers in April 2019 and on the Highland sleepers in October 2019. GBRf has a dedicated fleet of 10 92s (and two rescue locos) with dellner couplings and modifications capable of hauling MK5 sleeper stock on electrified routes and six 73/9s (sometimes piloted by a 66) on non-electrified routes.

Scottish Ministers agreed to terminate the franchise agreement with Serco Caledonian Sleeper on 25 June 2023 after Serco's proposal to 're-base' the franchise under new financial terms was rejected. A decision will be made in due course regarding the future of the Caledonian Sleeper with renationalisation and a direct award to Serco among options on the table.

In original Freightliner Grey, 90044 hauls a rake of MK2 daytime and MK3 sleeper coaches in the colours of its predecessor First Scotrail, working 1E43 21.43 Aberdeen and Fort William (two portions) to King's Cross at Harringay on 6 May 2019. Engineering works in the Euston area for HS2 resulted in the change of terminus with the sleeper portion from Inverness working as a separate train, due to platform length constraints at King's Cross.

After departing Edinburgh Waverley at 01.40 and finally nearing its destination at 07.30, CS-liveried 92010 powers 1M16 20.45 Inverness to Euston at Harrow & Wealdstone on 9 June 2022. This was the site of Britain's worst train crash in peacetime in which 112 people were killed and 340 injured on 8 October 1952.

92020, 028 and 043 carry the GBRf corporate colours despite being used on the Caledonian Sleeper. 92020 powers 1M16 20.45 Inverness to Euston sleeper at North Wembley on 2 June 2022. The 92 would work the service from Edinburgh Waverley where the portions from Aberdeen and Fort William join the Inverness portion, all hauled by a combination of 66/73 diesels.

Chiltern Railways

Chiltern Railways (CH), part of the Arriva Group, operates main line services on the Chiltern Main Line, along the M40 corridor between Birmingham Snow Hill and London Marylebone with peak time services extending to Kidderminster. Chiltern Railways also operates services on the London to Aylesbury Line with most services extended to Aylesbury Vale Parkway, which opened in December 2008. Both main lines are connected at each end of the Aylesbury to Princes Risborough branch line. Chiltern Railways also operates services to Oxford and Stratford-upon-Avon and local stopping services to Gerrards Cross as well as a Wednesday-only parliamentary train between West Ealing and West Ruislip, although this is showing as a bus service from December 2022.

Chiltern Railways, founded as M40 Trains Ltd, commenced operation of the Chiltern franchise when privatised from July 1996. CH inherited 34 BR Class 165/0s (these had displaced first generation Class 115 DMUs, which finished on the Chilterns in 1992) from Network SouthEast (NSE). In 2004, Chiltern received five Class 165s, which had previously transferred to Thames Trains back in 1993, thus operating the entire Class 165/0 fleet (two-car 165001–028 and three-car 165029–039).

Upon winning the franchise, new rolling stock was ordered with five new three-car (later extended to four-car) Class 168/0 (168001–005) Turbostars built between 1997–98 at Bombardier, Derby Litchurch Lane – the first rolling stock ordered by a TOC since privatisation in 1996.

Into the noughties, Chiltern's rolling stock was strengthened further with newly built Class 168s – with a bodyshell design resembling more of a Class 170. This included four-car 168106–107 and 168215–217, three-car 168108–113 and 168214/218 and 219.

From 2015, Chiltern Railways received nine two-car Class 170s (Nos 170301–170309) from First TransPennine Express, which were upgraded at Brush Loughborough to Chiltern's specifications and reassigned as Class 168/3 (168321–329).

From June 2011, to increase capacity, Chiltern ordered four two-car Class 172s (172101–104), although these could not be used over the London Underground routes as there was no place for the brackets to be mounted on their bogies for installation of tripcock safety equipment. These units were later leased in mid-2021 to West Midlands Trains (WMT) to bolster its fleet of 172 with no indication yet of a return to Chiltern.

As of 2022, apart from the GWR and Caledonian night-sleeper services, Chiltern is the only operator to use loco-hauled carriage sets (LHCS) in passenger services out of London, having acquired the rolling stock assets from ceased, open-access operator Wrexham and Shropshire Railway (an Arriva Company) in 2011.

From December 2014, Chiltern replaced DB Cargo 67s with newer Class 68s from Direct Rail Service (DRS) to work with MK3 carriages and a Driving Van Trailer (DVT), with three daily diagrams still operating in 2022. With running costs and noise pollution complaints from residents in London, it is unknown how much longer the LHCS have in service.

From the inception of the Chiltern Railways' franchise in 1996, there has been significant phasal investment in upgrading the route and services on the Chiltern Main Line. These projects, worked collaboratively by Chiltern, Network Rail and the DfT, were known as Evergreen initiatives and have drastically reduced journey times and increased service reliability and frequency.

CH was awarded a new six-year contract in December 2021, which runs until 2027, with a commitment to reducing emissions. HybridFLEX 168329 entered service in February 2022, fitted with a Rolls-Royce MTU hybrid drive system, meaning it can operate emission-free in stations when running on battery power.

From December 2014, Chiltern Railways started hiring Beacon Rail-owned Class 68s, which are leased to Direct Rail Services (DRS), to replace the older DB Cargo 67s. Built by Vossloh, 68010–015 were delivered from Spain in Chiltern silver/grey livery in 2013 and are used alongside DRS-liveried 68008 and 68009, although the latter has currently not got serviceable Electric Train Supply (ETS), thus restricting its usage. The 68s are fitted with Association of American Railroads (AAR) push-pull equipment so they can operate with MK3 carriages and a DVT. 68013 powers 1R33 13.09 Marylebone to Birmingham Moor Street at Northolt Park on 26 August 2017.

After working 1C06 06.20 service from Aylesbury Vale Parkway, 165003 and 165031 return with 5W10 07.31 Marylebone to High Wycombe ECS, appearing out of St John's Wood Tunnel at South Hampstead on 7 June 2022. The line then enters another tunnelled section called Hampstead Tunnel. This ECS was in readiness to form 2H16 08.09 High Wycombe to Marylebone peak service. This is one of two places where the Chiltern Main Line crosses the WCML and Watford DC line. In the shadows underneath are LNR 350250 and 350251 working 1Y04 05.33 Birmingham New Street to Euston service. 165003 was one of the five Thames Trains Turbos that returned to Chiltern in 2004.

The Chiltern Main Line out of Marylebone is the only main line out of London that is not electrified. 165038 works 2H21 10.33 Aylesbury to Marylebone stopping service crossing the WCML and Watford DC line between Sudbury & Harrow Road and Wembley Stadium on 9 October 2022. No Chiltern services call at Sudbury & Harrow Road or Sudbury Hill Harrow at weekends.

168002 leads 168108 working 1R36 14.00 Marylebone to Birmingham Moor Street along the 'new down fast' road (commissioned on 30 August 2011) near Northolt Junction, South Ruislip on 9 October 2022. On the right is the Suez Waste transfer station where West London household and domestic refuse is added to a daily containerised train from Brentford and then taken to Severnside, Bristol for incineration to produce electricity for the national grid. Previously, the Chiltern lines passed to each side of the waste transfer station until this section was remodelled under the 2010 Evergreen 3 project, with a new 100mph 'down' line constructed to run parallel with the 'up line', and the former 60mph 'down' line becoming a 'down loop line' used sparingly for trains stopping at South Ruislip station.

Four-car 168217 working 1H25 10.37 Birmingham Moor Street to Marylebone approaching Neasden South Junction, Wembley on 9 October 2022. Wembley Stadium station served by CH is the nearest station to Wembley Stadium, which can be seen in the background, albeit slightly obscured by new high-rise apartments. During stadium events, shuttle services can run between Marylebone and Wembley Stadium by utilising the turnback sidings just west of the station. To the east of the station, a new light-maintenance depot at Wembley was built in 2005 to service Chiltern's growing rolling stock. To the left is a freight terminal.

Two-car 168321 (formerly TP 170301) leads four-car 168215 working 1H29 11.37 Birmingham Moor Street to Marylebone at West Ruislip on 23 July 2022. In the background is the site of what will be the western portal of the new twin-bore 13.5km/8 1/2-mile Northolt tunnels for High Speed 2 (HS2), which will connect Old Oak Common station and West Ruislip, where the new HS2 lines surfaces. Two Tunnel Boring Machines (TBMs), named Sushila (launched on 6 October 2022) and Caroline, will dig towards Green Park Way (taking 22 months) and will eventually meet two other TBMs digging in the opposite direction from Victoria Road outside Old Oak Common. The quartet of boring machines will complete the Northolt tunnel, and the TBMs will be dismantled and lifted at Green Park Way (near Greenford) in what will become a ventilation shaft for the tunnels.

Chiltern Railways operates a Wednesdays-only parliamentary train between West Ealing and West Ruislip. This was previously a daily service from South Ruislip to London Paddington and return to High Wycombe, until the connection between the Great Western Main Line (GWML) and New North Main line (NNML), which runs from Old Oak Common to South Ruislip, was cut off for HS2 works. 165038 operated the last parliamentary train out of London Paddington on 7 December 2018. From 10 December 2018, it was re-routed to West Ealing via the Greenford Branch. The service was suspended during the Covid-19 pandemic and now runs empties every day except Wednesday, when it runs as 2M27 11.17 West Ealing to West Ruislip (ECS works in as 5M27 10.23 from Marylebone). From December 2022, the service is showing as a bus replacement. On 25 November 2021, 165013 works 5M36 14.00 Marylebone to West Ealing route refresher at South Greenford.

Chiltern Railways operates two trains per hour and an extra service in the peak in each direction over the London to Aylesbury line, which is shared with the London Underground Metropolitan Line between just south of Harrow-on-the-Hill and just north of Amersham. This section is owned and operated by London Underground, with tracks fitted with third and fourth rail electrical system and trainstop control. This means that Chiltern Class 165s and 168s, which operate over this route are fitted with trip-cocks on a beam at the cab end of the non-driving side. Two-car 165019 and three-car 165038 work 2C27 12.48 Aylesbury to Marylebone calling at Chorleywood station on 9 September 2017.

East Midlands Railway

Midland Main Line TOC (MML), owned by National Express, operated the Midland Main Line franchise from 28 April 1996 to 10 November 2007, inheriting HSTs from British Rail and ordering 16 four-car and seven nine-car Class 222 Meridian DEMUs (222001–023) in 2002. These new DEMUs, based on the 220/221 Voyager concept, were built by Bombardier in Bruges and first entered service on 31 May 2004.

After the DfT restructured the Midland Main Line franchise to incorporate East Midlands' services from Central Trains, Stagecoach won the newly formed East Midlands' franchise and commenced operations under the trading name East Midlands Trains (EMT) from 11 November 2007 until 17 August 2019.

The four 'Pioneer' Class 222/1s (222101–104), operated by Hull Trains, transferred to EMT during 2008/2009 bringing the Meridian fleet up to 27. Several Meridian formations have been altered under each franchisee to provide more capacity to current formations of four seven-cars (222001–004) and 23 five-cars (222005–023 and 101–104).

Abellio replaced Stagecoach from 18 August 2019 trading as East Midlands Railway (EMR), after Stagecoach submitted non-compliant bids.

EMR services are divided into three different brands, EMR InterCity services are long-distance services between St Pancras and destinations in the East Midlands and South Yorkshire using 222 Meridians and 180 Adelantes. Regional services for local operations across the East Midlands, using mainly DMUs and Connect services are electric suburban services between St Pancras and Corby using 360 Desiros.

For its InterCity operations, EMR inherited the 27 Meridians and repainted 222104 into its 'aubergine' colour scheme. The remaining 26 later received an 'outgoing' EMR livery.

EMR had inherited a fleet of 30 BR HST power cars; 24 of which were VP185-engined with nine eight-car sets. They were joined by six buffered and MTU-engine fitted power cars with three six-car sets, which had cascaded from Grand Central at the start of 2018. The final day of these HSTs in operation was 11 December 2020 with 43047 (the first to be fitted with a Paxman VP185 engine in 1995) and 43049 working 1F70, the 20.01 St Pancras to Leeds, the final Paxman VP185 HST in passenger service.

At the start of 2020, several London North Eastern Railway (LNER) HSTs were cascaded across to replace the original HSTs as they were more compliant with new accessibility regulations. However, their introduction was slow, hindered by the Covid-19 pandemic and significant corrosion repairs to the stock. They operated in reduced numbers in 2+6 formation before being withdrawn on 15 May 2021 – a year later than EMR had originally intended.

43274 was repainted into EMR purple and 43302 was rebranded with InterCity swallow livery to commemorate 39 years in service on the MML. This pair, the latter renumbered 43102, worked the final HST service, which was 1F70, the 20.01 St Pancras International to Leeds via Derby.

The acquisition of the Hull Trains' Class 180s (180109–111 and 113) in December 2020 enabled EMR to instantly retire its ex-GC 2+6 HST sets and allowed its remaining HST fleet, in mid-2021, to be retired, coinciding with the introduction of Class 360s on Corby services. EMR Class 180s work summer Saturday services to Skegness and InterCity services to Derby, Nottingham and Sheffield although their use is often limited because only Derby drivers sign them and often work in accordance with the maintenance schedules of the meridian fleet. They should be replaced by new Class 810s although their improved reliability means they may be retained for regional services.

EMR Class 360s (360101–121) entered service in May 2021 on the EMR Connect services between St Pancras to the recently electrified Corby. The fleet of 21 were cascaded from Greater Anglia during 2020/21 after being displaced by new Class 720s. 360102 and 112 received temporary vinyl wrapping of EMR colours for the launch at Bedford Cauldwell and Cricklewood respectively. All 360s have since been repainted at Arlington-Eastleigh, Hampshire.

Abellio will replace its EMR InterCity fleet of Class 222 and Class 180s with 33 new Hitachi five-car Class 810 bi-modal units, funded by Rock Rail and built at Newton Aycliffe.

On 10 August 2019, EMT commemorated 11 years of operating the East Midlands franchise by running a staff special covering many EMT main line routes using specially vinyl-wrapped power cars, 43081 and 43050. After joining the Gospel Oak to Barking line (known as the 'GOBLIN') at Junction Road Junction, Paxman VP185s, 43081 leads 43050 as they work 1Z25 08.02 Leeds to Derby service, passing the two portacabins that make up Upper Holloway signal box, before joining the East Coast Main Line via the Harringay curve. The special ran via the Midland Main Line (MML) to West Hampstead, taking the short hop across North London to the ECML before continuing to Grantham, across to Nottingham and finally to Derby via Sheffield. 43050 moved to Colas Rail while 43081 has entered preservation at the Crewe Heritage Centre.

One of only five former LNER/Virgin HST sets, which entered service with EMR instead of the originally planned nine, heads into St Pancras at the hands of 43272 and 43309 on the final day of HST operation on the MML on 15 May 2021. The PCs were to form 1D63 18.34 St Pancras to Leeds. These HSTs transferred from LNER in late 2020 having previously worked into Kings Cross, ironically located just a few yards across Pancras Road.

In February 2021, 43302 was renumbered 43102, repainted into InterCity swallow livery at Neville Hill and named *The Journey Shrinker*. It shares the world record with 43159 of the highest speed ever attained by a diesel when it ran at 148.5mph between Northallerton and York during test running of the SIG BT41 bogie on 1 November 1987. It is now preserved at the National Railway Museum, York. 43102 leads 43274 working 1D61 18.11 St Pancras to Nottingham at Childwickbury, near Harpenden on 18 April 2021.

222020, in East Midlands colours with EMR InterCity branding, departs St Pancras with 1F62 18.31 St Pancras International to Sheffield service on 15 May 2021. In Platform 1, 43309 and 43272 are ready to work 1D63 18.34 St Pancras to Leeds via Nottingham with the penultimate HST service from London. EMR services serving Leeds were cut during the 2022 summer-timetable change with 222001 working the last service from St Pancras (1F56) on 13 May 2022.

Departing from the 'cathedral of the railways', 222016 in the outgoing EMR livery, pulls away from St Pancras with 1F55 16.02 service to Derby. The Meridian had worked from Derby with 222009 as 1C56. However, 222009 was detached and worked 5F62 15.55 St Pancras circular via Kentish Town, returning in time to partner with 222014 on 1F62 17.32 to Derby. The future of the Meridians is unknown as they will be replaced by Hitachi A-family 810s from 2023.

Adelante 180s often work InterCity services at weekends to cover Meridians and bolster capacity. Carrying the 'outgoing' EMR livery, 180109 powers the 1D39 17.35 St Pancras International to Nottingham approaching Hendon station on 17 July 2022. The EMR 180s will be replaced with the arrival of Class 810 Aurora Bi-modes in 2023.

In the remnants of First Great Eastern/Anglia colours, 360105 leads EMR-liveried 360104 working 1Y29 16.44 St Pancras to Corby EMR Connect service under Brent Cross flyover, Staples Corner, Cricklewood on 17 July 2022. The following month, 360105 was repainted into new EMR colours at Arlington Eastleigh, Hampshire. Although the 360s have received repaints, they are yet to be refurbished inside and still have a 3+2 seating layout.

In new EMR colours, 360115 working 1Y04 06.11 Corby to St Pancras at Staples Corner on 8 August 2022. In addition to the IRA bombings, Staples Corner was infamous for the 1988 runaway of a pair of unattended Class 31s (Nos. 205 and 226) from the sidings at Cricklewood on the eastern side of the Midland Main Line, with both locos coming to rest on-top of each other on the westbound carriageway of the old A406 North Circular Road.

Elizabeth Line

The Elizabeth line (formerly TfL Rail) is operated by Hong Kong-headquartered MTR (Mass Transit Railway) Elizabeth line through a concession let by Transport for London. The Elizabeth line is a new east-west high frequency railway across London, which opened to passengers on 24 May 2022 by the opening of Crossrail's central section between Paddington and Abbey Wood.

The Elizabeth line was created by incorporating suburban commuter services on the GWML and the GEML and connecting them through Central London by new underground tunnels built by Crossrail Ltd. Passengers can travel from Reading or Heathrow in the west, through the London Core to either Shenfield, or via a new stretch of railway, to Abbey Wood in the east. Initially, Crossrail opened as three separate railways, but these were all integrated on 6 November 2022, significantly cutting journey times with a full timetable expected to start in May 2023.

In total, 70 new Bombardier Aventra Class 345s (345001–070) were built at Derby Litchurch Lane solely for the Elizabeth line, with 345005 the first to enter passenger service between Shenfield and Liverpool Street on 22 June 2017. The 345s entered service initially as seven-car trains due to platform-length constraints at Liverpool Street, although this was rectified during Easter 2021, as platforms 16 and 17 were both extended to accommodate nine-car trains. By the end of 2022, all 345s will have been extended to full length units.

TfL Rail was the interim brand that was used between the taking over of existing western and eastern services and the launching of through services using the Crossrail portals, eventually becoming the Elizabeth line. TfL Rail inherited 315818–861 from Greater Anglia when Shenfield to Liverpool Street metro services were absorbed into the Crossrail project on 31 May 2015. Since the introduction of new 345 EMUs in June 2017, the 315s have gradually been sent off lease with a handful retained for peak services until they finished in December 2022.

TfL Rail also inherited Heathrow Airport Holdings-owned Siemens Class 360s (360201–205) when the half-hourly 'Heathrow Connect' stopping services between London Paddington and Heathrow Airport were incorporated into the TfL network on 20 May 2018. Heathrow Airport Holdings ordered four Class 360s (360201–204) in 2003 from Angel Trains to work the (then new) Heathrow Connect services, which started on 12 June 2005. The Desiros were originally four-car units, later extended to five-cars in 2007. 360205 was ordered as an extra set, entering service in late 2006. It was re-liveried into Heathrow Express colours in 2010 to work the shuttle between Heathrow Central and Heathrow Terminal 4. This shuttle service ceased at the start of the Covid-19 pandemic.

There was a two-year delay in the introduction of the new Bombardier 345s on the Heathrow Airport branch due to software issues and compatibility with the ETCS signalling on the airport branch; 345004 was the first new Aventra to work to Heathrow Airport on 30 July 2020.

The 345s replaced the 360s on the Heathrow Connect services and they finished in September 2020. They have struggled to find work since, with two being sent for scrap.

TfL Rail took over the GWR half-hourly services between Hayes and Harlington and London Paddington with new Bombardier Aventra 345s commencing operations on 21 May 2018. This enabled the 12 GWR Class 387s to go to Ilford for modifications in readiness for the Heathrow Express. TfL Rail then took over the GWR-stopping services to Reading from 15 December 2019, which would be the precursor for the new Elizabeth line, eventually opening to passengers on 24 May 2022.

TfL Rail 315843 leads 315857 working 5C48 09.53 Liverpool Street to Gidea Park sidings at Stepney Green, near Bow Junction on 13 September 2018. The first 315 to be sent for scrap was 315850, which was delivered to CF Booths, Rotherham on 20 October 2018. A farewell running-day for the 315s between Shenfield and Liverpool Street took place on 26 November 2022.

The new 345 EMUs worked alongside the Class 360s on Heathrow services from July 2019 until they were removed from service on 13 September 2020. Heathrow Connect liveried 360203 with TfL Rail branding forms 2Y14 08.52 Heathrow Airport Terminal 4 to London Paddington arriving at Southall on 11 March 2019. The Southall Gas Works tower has since been demolished.

In Heathrow Express livery, 360205 powers 2T67 13.02 London Paddington to Heathrow Airport Terminal 4 at West Ealing on 29 February 2020. After finishing with TfL Rail, Rail Operations Group (ROG) purchased the five 360s from Heathrow Airport Holdings, storing them at Bicester MoD with the intention of returning them to traffic for logistics use. Unfortunately, technical issues and lack of interest resulted in 360205 being sent to Newport Docks for scrap on 23 August 2022 with 360204 following the next day. It was announced, in October 2022, that the Global Centre of Rail Excellence (GCRE) had purchased the remaining 360s (360201–203) for future use in research and innovation.

In early 2020, the 345s were extended from seven-car to nine-car trains on the GWML. Elizabeth line 345037 works 9P35 11.22 Heathrow Terminals 1, 2 and 3 to London Paddington at Acton Wells Junction on 13 June 2022. This is where the Dudding Hill freight line branches off from Acton Yard and the GWML.

Elizabeth line 345045 works 2W45 12.33 Shenfield to Liverpool Street service passing the Pudding Mill Lane Portal, west of Stratford on 29 September 2022. Passengers would need to change at Liverpool Street to access the core section of the Elizabeth line. With Crossrail fully open from 6 November 2022, trains now use the tunnel to enter the core section providing direct services between Shenfield and Paddington. From May 2023, some services from Shenfield will be extended further west.

Running parallel with the DLR, 345035 works 9U72 11.38 London Paddington XR to Abbey Wood XR arriving at the new Elizabeth line station at Custom House on 4 August 2022. This new section of line between London Paddington and Abbey Wood opened on 24 May 2022, with 12 trains operating per hour in each direction. From 6 November 2022, services were extended from Abbey Wood through to Heathrow Airport and Reading.

Eurostar

Eurostar operates around 20 services daily out of St Pancras to cross-channel destinations such as Paris, Brussels, Amsterdam, Disneyland, Paris and the French Alps.

Each Eurostar is formed of two half trains. This is to give each train the ability to split in two should an emergency arise in the Channel Tunnel. Eurostar currently has a fleet of Class 374s (374001–034), which are 34 eight-car half sets that form 17 full length trains. These were built by Siemens at Krefelds, Germany between 2011 and 2017. Eurostar E320 'Velaro' first entered service on 20 November 2015 working between St Pancras and Paris Nord and were the first train types with distributed power to work through the Channel Tunnel. They were introduced on services to Amsterdam from 2018.

These new second-generation Eurostars replaced most of the original Class 373 Eurostars, with just eight full sets (16 ten-car half-sets) having been refurbished and retained. The numbers are Class 373/0 Nos. 373007/08 and 015/016 owned by Eurostar UK and Class 373/2 Nos. 373205/06, 209/10, 211/12, 219/20, 221/22 and 229/30 owned by French Railways (SNCF). 3999 is a spare driving motor, which is based at Temple Mills, East London. All the other Eurostars have either been stored, scrapped, or preserved. These trains were the first to be built for cross-channel services for the opening of the Channel Tunnel in 1994.

Stratford International, despite its name, has never served an international market but is a great place to view passing Eurostars. Siemens 'Velaro' 374012 and 374011 working 9O24 12.31 St Pancras International to Paris Nord blast out of London Tunnel 1 and reach the gradient at Stratford International on 15 June 2022. Unlike the Class 373s, the lead vehicle carries passengers and the power pick up is mounted throughout the train.

The first-generation Eurostars were built by GEC Alsthom in the UK and France between 1992 and 1996 and across three different factories. The main assembly plant was at La Rochelle, France, and the power-car plant was in Belfort, France. Alstom Washwood Heath, Birmingham, England was the third plant, although some parts were also produced by La Brugeoise et Nivelles, Bruges, Belgium. In total, 31 Three-Capitals sets for Cross-Channel services were built with 373101 and 373102 the first full-length set to be released from Washwood Heath in 1993. 373001 and 373002 was the first Eurostar set built at Belfort, albeit in short formation, in 1992. Only eight original Eurostar sets remain in operation. In the UK, the Eurostar trains are now maintained at Temple Mills, East London. On 11 July 2022, 373230 and 373229 are working 9S47 12.02 Temple Mills Reception to St Pancras International awaiting a path on the Temple Mills chord at Stratford International. The E300 would form 9028 13.31 St Pancras to Paris Nord service. Each Eurostar has 18 carriages with two driving power cars at each end making them the longest trains to operate in the UK.

Grand Central

Grand Central (GC) is an open-access operator, which first launched in December 2007, connecting Yorkshire and the north east with London. Privately-owned Grand Central was bought out by Arriva in November 2011. Grand Central operates north-eastern services between Sunderland, Hartlepool and Kings Cross, and West Riding services between Bradford, Wakefield and Kings Cross.

Grand Central bid a fond farewell to its three HST sets on 31 December 2017 after ten years of service. These were replaced by Class 180 Nos. 180102–104, 106 and 108 from GWR. They joined 180101, 105, 107, 112 and 114 already with GC, creating a uniform fleet of ten Adelantes. Meanwhile, the HSTs transferred to EMT entering service in February 2018, enabling Class 222s to strengthen services elsewhere.

Currently, GC operate north-eastern services between Sunderland and Kings Cross with five services per day (four on Sundays) and West Riding services between Bradford and Kings Cross with four trains per day.

GC had planned to introduce five services between Blackpool North and Euston due to start in spring 2020 using DB Cargo Class 90s, MK4 rolling stock and a driving van trailer (DVT). This was abandoned due to the Covid-19 pandemic.

With just a few days left in service, Grand Central 43468 leads 43484 *Peter Fox 1942–2011 Platform 5* **working 1N90 16.50 Kings Cross to Sunderland service away from Kings Cross on 23 December 2017. The HSTs that worked the north-eastern services were replaced by 180s cascaded from Great Western Railway (GWR) following the introduction of new IEP trains.**

Having transferred from GWR to Grand Central in 2017, 180108 *William Shakespeare* works 1A83 12.11 Bradford Interchange to Kings Cross West Riding service at New Southgate on 24 July 2022. Grand Central introduced its West Riding service connecting West Yorkshire with London on 23 May 2010.

180101 with the driving motor car No. 59903 from 180103 leading, approaches Bowes Park station after using the turnback siding to work to Ferme Park via Kings Cross on 24 July 2022. The Adelante had a complete failure of all engines between Hornsey and Alexandra Palace working 1N95 15.23 Kings Cross to Sunderland service. After restarting two engines, 180101 limped to Alexandra Palace to detrain passengers before continuing to Ferme Park sidings. The DMU would work 5Z25 to Crofton Depot for repairs the following day.

Great Northern

Govia Thameslink Trains (GTR) took over the Great Northern and Thameslink franchises from First Capital Connect (FCC) on 14 September 2014, with both franchises incorporated into the new Thameslink, Southern and Great Northern franchise (TSGN).

Great Northern (GN) runs long-distance commuter services between Kings Cross, Peterborough, Ely and Kings Lynn and suburban commuter services between Moorgate, Welwyn Garden City and Stevenage. The operator inherited 44 1976/1977-built three-car Class 313s owned by Eversholt Rail (313018, 024–033, 035–064, 122–123 and 134) from FCC and its predecessor's livery was retained, with GN branding applied, through to their withdrawal in 2019. These worked the commuter services out of Moorgate.

The final Great Northern 313 service was the 2J83 23.33 Hertford North to Moorgate worked by 313048/049 on 30 September 2019 with 313064/134 working a charity farewell charter on 23 October 2019. All 44 examples have been scrapped with 313026 the first to leave Hornsey on 8 April 2019 and the last pair 313064/134 on 25 November 2019.

As part of the TSGN franchise agreement, GTR sought to replace its ageing 313 fleet. In February 2016, a deal worth more than £200m was finalised with Siemens to build 25 new six-car Class 717 (717001-025) EMUs at Krefeld, Germany. A variant of the 700 Desiro City Class, the new dual-voltage 717s provide 27 percent more capacity than 313s, are fully accessible, modern, spacious with air-conditioning, wi-fi and charging ports. 717005 and 717006 were the first to enter passenger service on 25 March 2019, with 717s taking over all ex-313 GN services on 1 October 2019.

The Class 365s were the last trains to be built at York Holgate Works between 1994 and 1995 and all 40 examples were inherited by GN at the start of the franchise. The Networker Express trains have gradually been phased out since 2017 by GN Class 387s and TL Class 700s entering service, although a handful were retained for the peak-busting Peterborough services.

The last remaining 365s were retired following the arrival of nine Class 387/2s cascaded from Gatwick Express (GX) with the last networker services operating on 15 May 2021, bringing 25 years of Class 365 services out of Kings Cross to an end. All 365s have since been scrapped apart from 365540, which is preserved at the East Kent Railway.

Built at Bombardier, Derby Litchurch Lane between 2014 and 2015, 29 four-car dual voltage Class 387/1s (387101–129) entered service between December 2014 and July 2015 as a stopgap on Thameslink (TL) services between Brighton and Bedford enabling the cascade of several Class 319s to Northern Rail and London Midland. When the Thameslink 700s entered traffic, the 387s were cascaded to GN in 2016 replacing Class 317s, 321s and relegating 365s to peak-hour services.

With GX services suspended during the Covid-19 pandemic and then on resumption, running a reduced service, nine surplus GX Class 387/2s (387201–209) transferred to GN in May 2021 to facilitate the retirement of the Class 365 'Networker Express' at the summer timetable change. The six former C2C 387s (387301–306) transferred to GN in July 2022 and entered service shortly afterwards, which enabled several GX 387s to move back to Southern and Gatwick Express.

The Class 313s inherited by Great Northern were looking worn in the months prior to their withdrawal. Still in the remnants of First Capital Connect colours, 313058 leads 313059 working 2K38 08.25 Welwyn Garden City to Moorgate out of Potters Bar Tunnel on 1 September 2018. Both units have since been scrapped.

With Arsenal's Emirates Stadium in the background, 313051 and 313053 work 2F70 13.48 Moorgate to Letchworth away from Drayton Park at the northern end of the Northern City Line and passing under the ECML at Clarence Yard Junction, Finsbury Park on 11 May 2018. The 313s would have just switched from 750 V DC third rail to 25 KV AC overheads at Drayton Park station.

In contrast to the Class 700 and 707s, the 717s leased by Rock Rail contain a fold-down emergency exit door at each cab end as per the regulations in the single bore Moorgate tunnels where there is no egress for side evacuation. 717017 works 2F08 16.10 Moorgate to Stevenage, taking the Hertford loop line at Wood Green South Junction, where it crosses over the ECML shortly after departing Alexandra Palace station on 29 September 2022. LNER 801103 can be seen on Bounds Green depot.

Nicknamed 'happy trains' due to their light cluster design and the installation of the cab air-cooling vent, during cab refurbishment in the early noughties, 365502 and 365520 work 1P85 09.10 Peterborough to Kings Cross service, exiting Hadley Wood north tunnels on 1 May 2021. From mid-2019, purple stripes were applied to each cab end of the 'networkers' to help prevent reflective glow for drivers when using driver-only operated platform monitors during dispatch.

The Electrostars work Great Northern services between Kings Cross, Ely and Kings Lynn as well as services to Peterborough in the same colour scheme as for Thameslink, but with Great Northern branding. 387128 leads 387122 working 1T09 06.44 Kings Lynn to Kings Cross at Welham Green on 1 May 2021.

Gatwick Express (GX) loanee 387204 leads 387119 working 1T10 07.42 Kings Cross to Kings Lynn at Harringay on 11 June 2022. These signals will disappear at some future point with the introduction of the European Train Control System (ETCS) as part of the East Coast digital programme between Kings Cross and Stoke Tunnel. Digital signalling helps calculate a safe maximum speed for each train to run, which will increase capacity and reduce delays, rather than being limited by signal block sections.

GWR and Heathrow Express

First Great Western (FGW) operate the Greater Western franchise with services out of London Paddington to the Thames Valley, Cotswolds, west of England, and South Wales. First Great Western has operated the franchise since 1998, after it bought out the shares in Great Western holdings owned by a group of British Rail managers (51 percent), FirstBus (24.5 percent) and 3i (24.5 percent), which ran the franchise since it commenced on 4 February 1996.

First Group rebranded as Great Western Railway (GWR) on 20 September 2015 in recognition of Brunel's legacy and the former Great Western railway, which operated between 1835 and 1947. The rebrand came at the start of a £7.5bn programme, which involved fleet upgrades including replacing ageing HSTs (built between 1975 and 1982 by BREL at Crewe) on long-distance services with new Hitachi bi-modal units, electrification of tracks and modernisation of stations. The final day of HST services out of London Paddington was on 18 May 2019 with four sets in service.

43002 was repainted into original Inter City 125 livery in May 2016 and 43185 repainted in Inter City swallow livery in August 2016 to commemorate the end of long-distance HSTs on the GWML and were star attractions at the now-closed Old Oak Common maintenance depot open day. The majority of HSTs finished their career with GWR in the FGW's blue dynamic lines livery with only several HSTs receiving the new GWR colours.

GWR retained a fleet of 14 HSTs trimmed down to four coaches and named 'Castle sets' to work secondary main line services in the west of England. GWR cascaded 54 power cars and more than 100 coaches to Scotrail for use on its Inter City services. Other power cars and coaches have since been preserved, stored (often for spares) or scrapped.

GWR has four Class 57s (57602–605) for its night Riviera Sleeper operations, with one Class 57 usually hired in from DRS for shunt movements. One sleeper train runs in each direction on the London Paddington to Penzance corridor every night except Saturday. GWR is considering loco replacements in the future as it looks to decarbonise its fleet.

GWR operate services in the Thames Valley, on the Windsor, Greenford, Henley and Harlow branches with Class 165 or 166 turbos. To provide greater capacity, more efficient and faster journeys on its suburban services, GWR leased 45 new four-car Class 387s Electrostars (387130–174) from Bombardier, delivered between 2016 and 2017, releasing the diesel Turbo fleet to other parts of the network. 387132 and 387131 were the first to enter service on 5 September 2016, running between London Paddington and Hayes and Harlington. Following further electrification, 387s started working to Didcot Parkway in January 2018 and to Newbury in January 2019. Since December 2021, the 387s have worked services to Cardiff Central and can provide additional capacity for events at the Principality Stadium.

As part of the Intercity Express Programme, GWR now operates 36 five-car and 21 nine-car Class 800s on long-distance InterCity Express services. All 800s were assembled at Newton Aycliffe, County Durham with the bodies being built and shipped from the Kasado plant in Japan. The 21 nine-car

800s were originally going to be Class 801 EMUs but were converted to bi-mode following delays in the electrification programme and reassigned Class 800/3. 800008 and 800009 became the first of the Hitachi AT300 family to enter service on the UK rail network on 16 October 2017. The remainder of the new fleet gradually followed leading to the withdrawal of HSTs in May 2019.

GWR also leases a fleet of 36 Class 802 IETs divided into 22 five-car Class 802/0 and 14 nine-car Class 802/1. Apart from 802001–002 and 802101, which were built in Kasado, Japan all were built between 2017 and 2018 at the Hitachi factory in Pistoia, Italy, with 802006/007 the first to enter revenue-earning service on 18 August 2018. Although Class 802s are identical to Class 800s, the former are better equipped to deal with the demands of the Devon and Cornish banks with higher-rated engine output, larger brake resistors and superior diesel range.

57605 *Totnes Castle* with 1C50 23.51 London Paddington to Penzance sleeper awaits departure from Brunel's iconic trainshed at London Paddington on 21 August 2022. Since the closure of Old Oak Common maintenance depot, the train is serviced at Reading TMD with maintenance carried out at Long Rock, Penzance. 57604 was on the rear, having hauled the train ECS from Reading where it would later detach. The MK3s used exclusively for the night riviera service are the only remaining slam-door stock in regular passenger operation on the UK rail network with GWR receiving dispensation from the Secretary of State for continued use despite not meeting 'Persons with Reduced Mobility' (PRM) requirements under the Railway Interoperability Regulations set out in 2011.

Named after and wearing a special commemorative livery celebrating the life of the longest and last surviving combat solider of World War One – Harry Patch, who passed away on 25 July 2009, aged 111. GWR 43172 leads 43124 working 1A12 09.47 Bristol Temple Meads to London Paddington service routed on the relief at Southall due to engineering works, on 20 May 2018. The 95m/311ft high and 60m/196ft diameter, redundant Southall MAN Gas Tower in the background was demolished the following year for property development. The 'LH' and upward arrow painted on its side shows the direction of London Heathrow Airport for air-pilot guidance. Above the lead power car is the iconic Grade-II listed brick water tower, completed in 1903 but disused in the '60s and now converted into residential housing. 43172 remains in traffic with GWR for its castle set operations but is now in GWR colours.

Engineering works on the GWML have, in the past, resulted in HSTs being diverted into Marylebone and/or Waterloo. Due to engineering works taking place between Slough and London Paddington over the Christmas period in December 2017, several GWR services were diverted via the Oxford-Bicester and Chiltern lines into Marylebone. On 24 December 2017, First Great Western (FGW) blue 43023 and 43028 arrive at Marylebone with 1Z22 09.28 service from Swansea.

On 18 May 2019, enthusiasts flocked to London Paddington to witness the final day of long-distance HST operation on the GWML. Four 125 HST sets were in service with all four being lined up parallel to each other with all sets departing London Paddington between 18.03 and 18.30. Working the penultimate HST service out of London Paddington, 43009 and 43002 pull away with 1W08 18.22 service to Hereford. Adjacent are 43198 and 43185 prepping to work the 1C26 18.30 service to Taunton – the final HST service out of London Paddington.

The 4km/2½-mile Slough to Windsor & Eton line has a three-trains-per-hour shuttle service using either a GWR Class 165 or 166. GWR two-car 165119 Turbo powers the 2A41 19.05 Windsor & Eton Central to Slough service away from Windsor at Eton Wick and over the impressive 1.8km/1.15-mile brick viaduct on 22 June 2022. The majestic Windsor Castle in the Berkshire countryside can be seen in the background.

GWR operates services on the Greenford branch using its turbo DMU fleet. With Crossrail taking over services between West Ealing and London Paddington, all services from Greenford terminate in the bay platform at West Ealing except for the first service from London Paddington and the last to London Paddington. In an effort for further decarbonisation, GWR is planning to trial a battery-powered Class 230 in 2023 with charging equipment installed at West Ealing station. 165125 working 2G42 15.52 West Ealing to Greenford arrives at Greenford sandwiched between the Central lines on 18 October 2022.

On Christmas Eve 2020, GWR 387171 and 387160 work 2P36 09.37 Didcot Parkway to London Paddington away from Hayes and Harlington station. When the 387s first entered service with GWR in September 2016, they only operated between Hayes and Harlington and London Paddington until the electrification of the GWML was extended west to Maidenhead in May 2017. Unfortunately, the overhead catenary makes clear photography of the GWML in London challenging.

Following the need to repair underframe cracks on the Hitachi 800 series units, from May 2021, GWR hired in three Class 387s (387301/02/06) from C2C to provide cover. A young enthusiast captures de-branded C2C 387306 leading GWR 387150 on 2P36 09.40 Didcot Parkway to London Paddington at West Ealing station on 11 April 2022. The West Ealing station signs have now been altered to incorporate the Elizabeth Line colours. The C2C 387s loan period with GWR ended in June 2022 with 37601 taking them to Nene Valley Sidings on 5 July 2022 prior to re-entering service with Great Northern.

An open day at Old Oak Common HST depot celebrated 111 years of operation on 2 September 2017 and raising more than £50,000 for children's charity 'Place to Be'. Many 'legends' of Great Western were on static display. An impressive line-up, from left to right of: Steam 6023 *King Edward II*, 7903 *Foremarke Hall*, D821 *Greyhound*, D1015 *Western Champion*, 50035 *Ark Royal*, 43002 *Sir Kenneth Grange*, 180102 and 800003 *Queen Elizabeth II/Queen Victoria*. 180102 and 180104, which had been used on Cotswold services, were on display but would shortly transfer to GC after being displaced by Class 800s. The final HST to leave Old Oak Common depot were 43093 and 43185 on 8 December 2018 before the site was demolished to make way for the new Old Oak Common HS2 interchange station due to open in 2026.

GWR 800319 working 1P21 09.59 Oxford to London Paddington passing Hayes and Harlington on 24 March 2022. In the background is the railhead-connected Hayes Asphalt Plant. The line between Didcot Parkway and Oxford was originally going to be electrified but was deferred due to increasing costs.

GWR five-car 802014 powers the 1D30 15.20 London Paddington to Oxford service and is about to pass the Hammersmith and City Line underground station at Westbourne Park on 26 July 2022. To the left of the frame is Paddington New Yard, where the new Crossrail lines are about to enter the western portal of the new Crossrail tunnels. A tarmac freight terminal is located under the 180m/200yd elevated bus deck, which hosts Westbourne Park Bus Station.

In the Pride month of June 2018, GWR celebrated diversity and inclusion and its support for the LGBT+ community by incorporating the 'Pride' flag and '#trainbow' branding into the GWR livery on 800008. Situated on Platform 8/9 at London Paddington, John Doubleday's bronze statue of railway engineer Isambard Kingdom Brunel, who designed London Paddington station and the Great Western railway, overlooks the arrival of 800008 *Alan Turing* in its trainshed having worked 1L96 20.19 from Swansea on 22 August 2022.

More than 200 HST coaches and 22 GWR Power cars were placed into store at Long Marston and Ely Papworth sidings following their retirement with GWR. After a period of storage with no new buyer or interest found, Angel Trains started sending HST trailers and Power cars for scrapping at Eastleigh and Newport Docks. Operated by ROG, hired-in Harry Needle 20311 leads 20314 and 47813 working 5O86 09.32 Ely Papworth sidings to Eastleigh approaching Chertsey on 20 April 2020. The ex-GWR HST trailers are sandwiched between the barrier vehicles.

Heathrow Express

Heathrow Express (HEX) is an open-access operator with an airport shuttle service between London Paddington and Heathrow Airport. Originally, it was operated solely by Heathrow Airport holdings when it commenced service on 23 June 1998. The traction was nine four-car (332001–004, 010–014), and five five-car (332005–009) Class 332 EMUs, built by CAF at Zaragoza, Spain between 1997 and 1998 with Siemens traction equipment installed. These worked for HEX until their withdrawal in December 2020. From August 2018, and running until at least 2028, the operational management of HEX trains was incorporated into the Greater Western franchise, although Heathrow Airport would still own the service, manage its stations and commercial aspects.

GWR would introduce 12 four-car Class 387s (387130–141) dedicated to Heathrow Express services from December 2020 and replace the 332 fleet. This nullified the need to build a new depot at Langley for its predecessor as the HEX depot at Old Oak Common had to be vacated for HS2 works and the 387s could be maintained at Reading.

With no train protection and warning system (TPWS), corrosion issues and considerable attention required to make them compatible with other areas of the UK rail-network, HEX sent its units for scrap. 332014 was first to depart for Sims Metal in Peterborough in November 2020, with the rest being scrapped at Newport Docks by mid-February 2021. However, three cars from 332001 were preserved at Siemens factory in Goole.

The 12 387s received modifications and a repaint at Ilford for use on the Heathrow Express airport shuttles. This included upgrading the interiors to first class, installation of wi-fi, additional luggage space and ETCS signalling equipment for use in the Heathrow airport tunnels.

With only two days left in service, 332010 prepares to work 1Y62 15.12 service to London Paddington at Heathrow Terminal 5 on Boxing Day, 2020. 332010 and 332007 were taken to Newport Docks for scrapping 17 days later.

Right: Unusually routed on the 'up' slow due to engineering works, 332013 works 1Y36 08.42 Heathrow Airport Terminal 5 to London Paddington at Hanwell on 17 March 2019. 332013 and 332006 were hauled to Newport Docks for scrapping on 2 February 2021.

Below: 387131 and 387137 work 1T52 12.10 London Paddington to Heathrow Airport Terminal 5 over the Grade-I listed Wharncliffe Viaduct, Hanwell on 7 February 2022. This was the first significant structure designed by Isambard Kingdom Brunel and the first to carry electrical telegraph cables.

Greater Anglia

The East Anglian franchise commenced on 1 April 2004 through the merging of separate franchises. The Great Eastern franchise operated by First Great Eastern were commuter services between Liverpool Street and Ipswich including the Essex branches. The InterCity Anglia franchise operated by GB Railways, trading as Anglia Railways, were InterCity services between Liverpool Street and Norwich and Anglian regional services. The final piece was the West Anglia segment of the National Express-operated Great Northern West Anglia franchise, which were services between Liverpool Street, Stansted, and Cambridge. National Express East Anglia (NXEA), originally trading as ONE, held the franchise from its inception until Greater Anglia, owned by Abellio, took over from 5 February 2012. Abellio later sold 40 percent of the business to Mitsui & Co Ltd.

When Abellio retained the franchise when retendered in October 2016, it was committed to undertake a complete rolling stock reform, with 30 Renatus 'Dusty Bins', which are set to leave in early 2023, currently the last remaining of its legacy fleet.

Greater Anglia (GA) has replaced its 15 BREL Class 90 fleet with ten 12-car Class 745 (745001–010) Stadler FLIRTs built in Bussnang, Switzerland between 2018 and 2020. They entered service in January 2020. The final LHCS on the Great Eastern Main Line (GEML) was 90001 working the 17.00 Norwich to Liverpool Street and 19.30 return. The 90s originally entered service on the WCML in 1988 but were cascaded from VT to Anglia following the introduction of Pendolinos from 2002 and replaced the Class 86s.

The 317/1s, Nos. 317301–317348, were built between 1980 and 1982 and a second batch, which were assigned 317/2s Nos. 317349–317372, were built between 1985 and 1987. The MK3-based EMUs endured a peripatetic life until 2004 when they all started working out of Liverpool Street on Anglia suburban services for ONE, (all later inherited by Greater Anglia,) apart from 317337–348, all of which worked GN services for FCC (followed by GTR Thameslink) until cascading to Greater Anglia in 2017.

Apart from the 12 Great Northern 317/1s, all 317s have been renumbered through various refurbishment programmes. The 317s were eventually withdrawn from service in July 2022, having surprisingly outlasted the 379s on West Anglian services.

The Class 321s, nicknamed Dusty Bins after an animated character on the television show *3-2-1*, offered some nice colour variations on the GEML. They were built by BREL at Doncaster Holgate Road between 1987 and 1991. A batch of 66 321/3s (321301–366) was ordered originally for NSE services to Cambridge and for the Essex branches out of Liverpool Street, replacing slam-doors Classes 305/7/8 and 9. A follow-on order of 48 321/4s (321401–448) with more first-class seating followed for suburban services on the WCML. The 321/4s have experienced a more nomadic life with operators such as Silverlink, London Midland, First Capital Connect and Great Eastern, and all finished working in East Anglia or were converted to a Class 320 for Scotrail. At the end of 2018, prior to the arrival of the 720s, Greater Anglia had all 66 321/3s and 35 321/4s in service. Several 321/4s worked for GA before leaving for Scotrail.

All the non-Renatus 321s have left Greater Anglia with 321357 and 321446 the first to go for scrap on 26 January 2021. The rest are either stored, possibly awaiting future use (such as conversion to a Class 600 hydrogen unit) or have been scrapped.

The 21 four-car 100mph Siemens Class 360s (360101–121) replaced the outgoing slam-door Class 312s from August 2003 and worked commuter services on the southern section of the GEML between Liverpool Street and Ipswich including the electrified Essex branch lines until the entire fleet were transferred to EMR in 2020/2021. Due to delays with the introduction of the new 720 units, the fleet's departure to EMR was aided by the transfer of three Class 321/9s and five Class 322s from Northern.

Greater Anglia inherited a fleet of 30 four-car Class 379s (379001–030), which had been built for NXEA between 2010 and 2011 at Bombardier, Derby Litchurch Lane to increase capacity on West Anglia routes and replace ageing 317s. Ironically, the cheaper to lease 317s outlasted the 379s working for GA and a new operator is being sought.

GA ordered ten 12-car Class 745/1 (745101–110) Stadler FLIRTs for the Stansted route to replace the 379s with the first 745/1 entering service on this route on 28 July 2022. GA also ordered 22 ten-car (720101–122) and 89 five-car Class 720s (720501–589) on 29 September 2016 from Bombardier (later Alstom) for its suburban services out of Liverpool Street to replace its Class 317, 321, 322, 360 and 379 fleets. This order was later amended in autumn 2020, cancelling its ten-car fleet, and increasing its five-car 720 fleet order to 133 for greater flexibility, albeit retaining the 720/1 numbering (720101–144). Following software glitches and the Covid-19 pandemic delaying their introduction originally forecasted for March 2019, 720515 and 720517 finally entered service initially on Southend services on 26 November 2020. The delay in their arrival forced GA to seek dispensation from the DfT and retain EMUs that did not meet accessibility requirements into 2022.

In the London Borough of Tower Hamlets, 90005 *Vice Admiral Lord Nelson* **hauls 1P31 11.00 Norwich to Liverpool Street service at Stepney Green near Bethnal Green on 18 September 2019. Due to the Covid-19 pandemic, there was no official farewell rail tour to commemorate the last LHCS on the GEML. 90001 and 90002 moved to charter operator Locomotive Services Ltd and 90003–15 were transferred to Freightliner, to work container trains between Scotland and East Anglia.**

The electrified Canonbury chord connects the ECML at Finsbury Park to the North London line at Canonbury and is used for freight and stock movements. From spring 2018 until withdrawal in early 2020, one GA MK3 set would work via the chord to Bounds Green for daily maintenance instead of stabling at Orient Way sidings. This was due to limited space at Norwich Crown Point for maintenance, as modernisation works were taking place for the introduction of the new Stadler fleet. With Arsenal's Emirates Stadium to the left, 90008 *The East Anglian* hauls 5L54 16.54 Bounds Green TMD to Liverpool Street passing the narrow island platforms of Drayton Park station on the Northern City Line on 1 August 2018. The greenery on the left side was the site of the former Northern Line Highbury branch stabling sidings, which were made redundant when the line ceased to be part of the London Underground network in October 1975.

All Class 317/2s became 317/6s with 300 added to their original number during refurbishment at Wolverton works in 1998–99. 317661 (ex-317361) leads 317658 (ex-317358) working 2H19 09.06 Cambridge North to Liverpool Street across the River Lee Navigation, Clapton on 13 March 2018. All 21 Class 317/6s were scrapped at Eastleigh works during 2021.

With delays to the introduction of the Class 720s Aventra units because of the Covid-19 pandemic and software issues, as well as the withdrawal of the Class 379 fleet in February 2022, the GA 317s were still running into May 2022, albeit only for cover and contingency at this point. As indicated by the light blue doors, ex-GN 317343 and 317338 power 1B91 16.57 Stansted Airport to Liverpool Street over the River Lee Navigation, Clapton on 5 July 2022. 317338 and 317339 were the last two 317s to move to Ely Papworth sidings for storage, being dragged by 37884 from Cambridge TRSMD on 28 July 2022.

In total, 133 five-car Aventra Class 720s are replacing the entire Greater Anglia Class 317/321/322/360 and 379 fleets. Facing imminent retirement, 317338 and 317343 work 1B72 13.40 Liverpool Street to Stansted Airport passing 720559 and 720575 working 2F39 12.35 Colchester Town to Liverpool Street at Bethnal Bank on 5 July 2022. London Overground 378144 is crossing the GEML towards Shoreditch High Street with 9F31 13.27 New Cross to Dalston Junction.

317343 and 317508 were scheduled to run nine return trips between Liverpool Street and Hertford East to give enthusiasts their final chance to bid farewell to the class on 16 July 2022. Carrying a farewell headboard, ex-GN 317343 leads GA-liveried 317508 (renumbered from 317311 during refurbishment by ONE in 2005) working 2O02 18.12 Liverpool Street to Hertford East, near Cambridge Heath on 16 July 2022. Although the 317s were on standby for a short time afterwards, this did turn out to be their last appearance in passenger service.

321426 and 321445 work 5M24 09.57 Liverpool Street to Orient Way sidings at Stepney Green east of Bethnal Green on 13 September 2018. 321426 is in former National Express East Anglia livery, now rebranded as Greater Anglia. National Express operated the East Anglia franchise from April 2004 to February 2012 (although originally traded under the name ONE until February 2008), when the franchise was discontinued and awarded to Abellio. 321445 has been scrapped at CF Booths, Rotherham and 321426 is currently stored at Worksop.

Ex-GN Dusty Bin, wearing the remnants of First Capital Connect colours, 321418 works 2K18 11.14 Liverpool Street to Shenfield at Stepney Green on 2 September 2018. Following the arrival of 387s from Thameslink to Great Northern in 2016, 321418 was one of several FCC 321s cascaded to Anglia. 321418 would later be dragged by 37884 to Kilmarnock depot on 11 November 2018 to be converted into a Scotrail Class 320 with one trailer being removed to form a three-car set.

Between 2016 and 2018, as part of the Renatus programme, 30 Class 321s (321301–330) were fitted with a new Vossloh Kiepe traction package, a completely rebuilt interior including installation of air conditioning, new seats, energy-efficient LED lighting, larger vestibule areas, wi-fi, plug sockets and a PRM (person with reduced mobility)-compliant accessible toilet. Despite such investments, GA will replace all its overhauled 321s with new 720s, although they will be the last to go off lease. Renatus 321325 and 321317 work 1Y16 12.02 Liverpool Street to Ipswich service crossing Grove Road, Mile End on 2 June 2022.

Class 360s carried an adaptation of their original First Great Eastern colours, with 'Barbie' stripes and 'First' vinyls removed and with National Express East Anglia-branded white stripes added from the time it ran the franchise. No Class 360s were painted in GA livery, the only change being the branding change to 'Greater Anglia'. 360107 works 1Y12 10.02 Liverpool Street to Ipswich service at Manor Park station on 2 October 2019. All 360s later transferred to EMR.

In mid-2020, to help release Class 360s to EMR following the slow introduction of the 720s, surplus Class 321/9s (321901–903) and Class 322/4s (332481–485) transferred from Northern to Greater Anglia. 322482 leads 321323 working 1N18 10.18 Liverpool Street to Clacton on Sea at Bethnal Green on 5 July 2022. All the ex-Northern Dusty Bins have since been scrapped at Newport Docks.

Introduced in 2011, the 379s operated on Cambridge, Ely, Hertford East and Stansted Airport suburban commuter services and Stansted Express services alongside Class 317s. GA replaced the 379s with new Class 720s and 745/1s. Due to high leasing costs, they were all put into store (mainly at Harwich) in early 2022 pending future use with a new operator. 379017 leads 379009 working 1B09 09.24 Stansted Airport to Liverpool Street at Turnford Brook, Cheshunt on 15 May 2020.

The 745 FLIRTs were the first brand new trains introduced on the GEML for more than 60 years. After delays entering traffic, the first 745 entered service on 8 January 2020 and the fleet had replaced all the Class 90 LHCS by the end of March 2020. Swiss-built 745006 powers the 1P22 11.02 Liverpool Street to Norwich service over Regent's canal at Mile End on 2 June 2022.

Although built for Stansted Express services, the Class 745/1s were introduced and can be found working alongside Class 745/0s on the GEML InterCity services and vice versa. Four new GA trains pass each other on Bethnal Bank just outside Liverpool Street on 5 July 2022. On the West Anglia Main Line, Stadler 745108 approaches Bethnal Green with 1B58 11.55 Liverpool Street to Stansted Airport, passing 745103 the 1B45 11.12 Stansted Airport to Liverpool Street. On the Great Eastern Main Line, Aventra 720539 and 720554 work 1K40 11.55 Liverpool Street to Southend Victoria, passing 720545 and 720553 on 1F39 11.00 Braintree to Liverpool Street.

More of Greater Anglia's legacy fleet was withdrawn as more Class 720s entered service. Built at Derby Litchurch Lane by Bombardier (later Alstom), 720581 and 720525 work 1F41 12.00 Braintree to Liverpool Street west of Bow Junction, Stratford on 2 June 2022.

In July 2022, the Class 317s were withdrawn meaning that Class 720s had replaced all 317s and 379s on West Anglia and Lea Valley services. The arrival of the last remaining 720s is expected to be in mid-2023. 720517 passes Temple Mills Eurostar depot working 2M31 14.08 Meridian Water to Stratford service on 15 June 2022. Eurostar shunter 08948 is next to stored Eurostar set 373215 and 373216, which is currently split in half.

Hull Trains

Owned by First Group, open-access operator Hull Trains (HT) operates services between Hull, Beverley, and Kings Cross. It was established initially by GB Railways (which was later acquired by First Group in 2003) and two former British Rail managers. Hull Trains launched its first service between Kings Cross and Hull on 20 September 2000, originally using Class 170 Turbostars. In 2005, four four-car Class 222 Meridians (222101–104) replaced the Class 170s with six daily services. In 2008, Hull Trains acquired four five-car Class 180s leased from Angel Trains (180109, 180110, 180111 and 180113) to replace the meridians, which were cascaded to East Midlands. The 180s worked for Hull Trains until 2020 when they were replaced by new Class 802 units. However, there were reliability issues with the 180s, which resulted in GWR HSTs being hired-in, made surplus by GWR following the introduction of IETs (InterCity Express Trains).

In March 2016, Office of Rail and Road (OPR) announced it had approved an extension of track access for First Hull Trains to at least 2029 (extended in 2021 for another three years to 2032). This led to Hull Trains proceeding with a £60m deal with Hitachi to build five new five-car Class 802 (802301–305) bi-modals to replace its Diesel Adelante fleet. HT currently operates five daily return services between Kings Cross and Hull and two between Beverley and Kings Cross on weekdays, with one return service at weekends.

In First Group neon blue livery, 180113 powers 1H04 13.48 Kings Cross to Hull service, as it approaches Potters Bar Tunnel on 17 January 2019. The four 180s would later be cascaded in 2020 to EMR, with 180111 moving to Derby Etches Park on 7 January 2020. They entered service with EMR at the 2020 December timetable change replacing the former GC HST sets.

During 2018, the Alstom Coradia Class 180s were plagued by reliability issues, which led to Hull Trains hiring a GWR HST formed of five coaches in January 2019. This was followed by an additional HST in April 2019. The HSTs would provide cover until the arrival of the Paragon fleet in December 2019. FGW (First Great Western) blue-liveried 43190 and 43010 arrive at Kings Cross with 1A94 13.31 Hull to Kings Cross on 9 March 2019. DB Cargo 67010 is on LNER Thunderbird duties.

Nicknamed 'paragons' after Hull station, 802301 was the first paragon to enter passenger service on 5 December 2019. These new AT300s allowed the hired-in HSTs to be returned to GWR and the 180s to be cascaded to EMR. Hull trains suspended its services because of the Covid-19 pandemic during each of the three UK lockdowns but resumed services again on 12 April 2021. 802303 powers 1H02 09.48 Kings Cross to Hull at Holloway Road on 9 June 2022.

Lumo

Lumo is a new open-access operator owned by First Group, offering cheaper and alternative rail services on the East Coast Main Line (ECML) between Kings Cross and Edinburgh with trains calling at the stations of Stevenage, Newcastle, and Morpeth. Services commenced on 25 October 2021 using Beacon Rail-financed Hitachi five-car Class 803s, Nos. 803001–005. Although they are like Class 801s, these EMUs do not have a diesel engine to power the train in the event of failure or wires down, but a battery to continue powering onboard facilities. Lumo services went up to ten trains per day in 2022.

On 11 June 2022, 803002 works 5S93 09.15 Ferme Park to Kings Cross heading south around the back of Harringay station after using the turnback siding at Bowes Park. This unit then formed 1S93 10.25 Kings Cross to Edinburgh service.

LNER

The InterCity East Coast franchise was nationalised under the name East Coast in November 2009 after National Express ran into financial difficulty despite operating the franchise for less than two years. Its predecessor Sea Containers, operating as GNER, likewise ran into financial difficulty and was stripped of the franchise, having operated the franchise from its inception in April 1996.

The franchise was re-privatised from 1 March 2015 with a joint venture between Virgin Trains (10 percent) and Stagecoach (90 percent) being the successful bidder to operate the franchise for eight years. This contract was subsequently terminated early in mid-2018, again due to financial problems because of over-bidding and a breach of its £3.3bn terms agreed with the DfT resulting in the franchise being brought back into public ownership. Following the stripping of Virgin Trains East Coast (VTEC) from the franchise, London North Eastern Railway (LNER), a DfT company owned by the government, commenced operation of the InterCity East Coast franchise from 24 June 2018 with all its legacy rolling stock being rebranded.

LNER inherited 31 Class 91s (91101–22, 24–32), which worked with MK4s on long-distance electrified services between London and Scotland on the East Coast Main Line (ECML) until they were gradually replaced by new Azumas as part of the DfT's InterCity Express Programme in 2019–20. The 91/0s were built by Adtranz Bombardier at Doncaster between 1988–91. After refurbishment in 2000–02, they were reclassified into the Class 91/1 series. 91023 was renumbered as 91132 as it was considered 'unlucky' after being involved in both Selby and Hatfield rail crashes. 91110 and 91111 both carry commemorative war-time liveries, and the former holds the British speed record for a locomotive travelling 260kmh/162mph at Stoke Bank, between Peterborough and Grantham, on 17 September 1989. The 91s were being withdrawn from July 2019 and several have already been scrapped.

VTEC, later LNER, also hired in Class 90s to cover for 91s on maintenance from 2016. Initially LNER was set to withdraw its remaining IC225 fleet by the end of June 2020 but opted to retain ten sets to fulfil its franchising timetable commitments and to provide cover for its Hitachi trains to receive repairs. The Electras work between three and five daily diagrams between London and West Yorkshire.

Introduced on the ECML from 1978 and carrying the baton forward from the popular Deltic locomotives, the Class 43 High Speed Trains became known as 'the journey shrinker' and were a game-changer in providing faster and more reliable services for BR. LNER inherited 32 Class 43 power cars from VTEC and 14 HST sets comprising of nine coaches, not including set NL55 on loan from East Midlands. With the acceleration in the arrival of new Hitachi-built Azumas, LNER was able to withdraw its HST fleet by the end of December 2019.

LNER repainted 43206 and 43312 (renumbered 43006 and 43112) and seven coaches into original 1970s Inter-City blue/yellow livery for a four-day farewell charter across the ECML and Scotland, raising money for the charity CALM between 18 and 21 December 2019. This brought the HST (originally classified as Class 254 by BR) work on the ECML to a close after four decades and changing the experience of rail travel for passengers forever. Having been refurbished in 2016, nine HSTs sets transferred to EMR to replace its existing HST fleet as they were more compliant with PRM regulations until withdrawal in May 2021.

As part of the DfT's £5.7bn InterCity Express Programme to replace the ageing HSTs and Electra 225 sets, 23 bi-modal 800s and 42 Electric 801s were built at Kasado, Japan and assembled at Newton Aycliffe depot between 2014 and 2018. LNER's fleet of 23 Bi-modal Class 800s are made up of 13 nine-car 800/1 sets (800101–113), and ten five-car 800/2 sets (800201–210). These transverse non-electrified routes and have replaced the HST sets as well as some Class 91 diagrams.

LNER's fleet of 42 Electric Class 801s are made up of 12 five-car 801/1 sets (801101–112), and 30 nine-car 801/2 sets (801201–230) as a direct replacement for the Class 91s.

The Azumas were meant to initially enter service with LNER in December 2018. However, electro-magnetic emission issues with lineside equipment meant their introduction was delayed until 15 May 2019, when the first LNER 800s entered service working initially between Leeds, Hull and Kings Cross. 801101 was the first Class 801 to work on the UK rail network on test in January 2018, although it wasn't until 16 September 2019 when the Class 801 made its debut in passenger service, with five-cars 801109 and 801110 paired together on Leeds and Newark diagrams. The first nine-car 801/2s entered service on 18 November 2019, which aided the start of withdrawing HSTs from service. All 42 Class 801s had entered service within a year.

An unusual scene in London as 91124 in Virgin Trains East Coast (VTEC) livery is showered in a blizzard from Anticyclone Hartmut as it departs with 1D20 15.35 Kings Cross to Leeds on 2 March 2018. 91124 was the Electra used to debut the new VTEC colours exactly three years earlier.

91105 working 1D26 18.03 Kings Cross to Skipton passing GN 313122 and 313046 (both now scrapped) working 2F26 17.50 Moorgate to Watton-at-Stone at Hornsey station on 6 June 2019. With the first Azuma entering service in May 2019, LNER started to withdraw its Electra fleet with 91108 the first to be withdrawn on 22 July 2019 after accumulating 12,231000km/7.6 million miles since entering service.

As part of essential maintenance at Doncaster Wabtec, LNER repainted its IC 225 sets into a new livery, a hybrid of the classic InterCity livery and the new LNER Azuma 'oxblood' colours. The first Electra to carry the new colour scheme was 91127 in June 2022. On 8 August 2022, 91127 powers 1D10 10.33 Kings Cross to Leeds at Holloway Road. The IC225 sets are likely to stay until around 2025 with newly built CAF units suggested replacements.

From late 2016, VTEC hired up to three Class 90s from DB Cargo for use on services between London, Leeds, Newark, and York and occasionally Newcastle due to locomotive shortages. On 6 June 2019, at Alexandra Palace 90019 powers 1B88 16.06 Kings Cross to Newark North Gate. Alexandra Palace station was formerly known as Wood Green until its name was changed in May 1982. The last 90 diagrams were on 14 June 2019 following the arrival of new Azuma rolling stock.

For several years, East Midlands Trains would provide power cars and an HST rake (sometimes multiple) as cover for stock shortages on the ECML. It was also not unusual for the spare Cross Country HST at Craigentinny, near Edinburgh, to be used. On 23 December 2017, 43075 and 43061, with set NL55 on hire to LNER, await departure from Kings Cross with 1H10 17.10 service to Leeds. The former EMT set would be the first to go off lease in November 2019. 43061 and 43075 were both taken to Newport Docks for scrapping on 27 October 2021.

Exiting the twin-bore tunnels at Potters Bar, LNER 43239 leads 43290 working 1A04 06.05 Leeds to Kings Cross service, on 13 May 2019. 43239 worked the last HST-operated *Highland Chieftain* on 9 December 2019 (with 43296), and today is in service with Arriva Cross Country along with 43208. 43290 continued with LNER and is now with Colas Rail, often deployed on test train duties.

LNER started withdrawing its HST fleet from November 18, 2019, with its final day in service on 15 December 2019. With just two days to go, EMR-bound 43238 and 43316 prepare to work 1B90 19.06 service to Lincoln Central at Kings Cross on 13 December 2019. After working for EMR, several of the former LNER power cars moved to Ely Papworth sidings for storage including 43238 and 43316.

The new Intercity trains on the East Coast were named 'Azuma' by VTEC meaning 'East' in Japanese. On 8 August 2022, five-car bi-mode 800202 works 1A12 06.58 Hull to Kings Cross, passing the Emirates Stadium at Holloway Road.

LNER 800104 carries celebratory Scots branding and a tartan on its DTFO (Driving Trailer First Open) and was used to launch the new Azuma Anglo-Scots services. It travelled from Edinburgh working the 05.40 *Flying Scotsman* service on 1 August 2019, to Aberdeen on 25 November 2019 and the *Highland Chieftain* to Inverness on 10 December 2019. The introduction of electric Class 801s entering service in September 2019, enabled the movement of nine-car bi-modal Class 800/1s to Anglo-Scot routes replacing HSTs. With its tartan branding at the back, 800104 works 1D08 09.33 Kings Cross to Leeds service, having exited Copenhagen Tunnel at Holloway Bank on 14 June 2022.

Owned by Agility Trains, the Class 801s (despite being classified as an EMU) do have a small diesel engine for emergency and auxiliary use, which can also be utilised when paired with a Class 800/2. 801207 working 1E12 11.00 Edinburgh to Kings Cross at Hornsey on 29 September 2022.

London Northwestern Railway

Owned by Govia, London Midland (LM) ran the West Midlands franchise from 11 November 2007 to 10 December 2017. This was a franchise created through the merging of Central Trains and Silverlink operations and included express and suburban commuter services between Euston and the West Midlands. London Midland eventually had a fleet of 77 four-car Class 350 Desiros.

It inherited 30 Class 350/1 (350101–130) from Central Trains/Silverlink, and as part of the franchise commitment LM ordered 37 Siemens Class 350/2s (350231–267) in November 2007, which entered service in December 2008 to replace the 321s, which were cascaded to other rail operators. A further ten Desiros Class 350/3s (350368–377) were built and entered service from October 2014 to enhance capacity.

West Midlands Trains (WMT, a consortium initially of Abellio, East Japan Railway Company and Mitsui & Co Ltd) won the West Midlands franchise from 11 December 2017 and operated services between London and the West Midlands under the trade name London Northwestern Railway (LNR) with the franchise expected to run until at least March 2026.

The Class 350 fleet was bolstered to 87 examples with the arrival of 10 TransPennine Express Class 350/4s (350401–410) in 2019/2020 after they were displaced by 12 new Class 397s. All LNR's Siemens Desiros are owned by Angel Trains, apart from the Class 350/2 fleet, which are owned by Porterbrook. LNR's stalwart fleet of 37 Class 350/2s will be returned to Porterbrook due to high leasing costs and as part of its franchise commitments to introduce new trains and increase capacity. The replacements will be 36 new LNR Aventra five-car Class 730s, which are expected to enter service in 2023, although 48 three-car 730s bound for West Midlands Railway (WMR) will replace the 319s initially. These are currently being built at Derby Litchurch Lane by Alstom (formerly Bombardier).

Displaced by Thameslink Class 387s and Class 700 EMUs, seven 1987–88-built Class 319s transferred from Thameslink to LM in 2015 for use on the Watford–Albans Abbey branch and suburban peak services in and out of Euston. These replaced the seven Abellio Scotrail-bound Class 321s (321411–417) that had been retained for these services. Since WMT took over the West Midlands franchise in December 2017, more surplus 319s arrived bringing the fleet total up to 15. This enabled WMR to take several 350s out of service for planned refurbishment, with every Desiro receiving an interior refresh. The former Thameslink 319s are expected to continue with LNR on peak services in and out of Euston until January 2023, when they will be replaced initially by new three-car Class 730 Aventra EMUs and retired from service. Several have already gone off lease and subsequently been scrapped.

A Class 350 had already replaced the Class 319 on the St Albans Abbey Branch diagram in 2021 and the summer timetable in 2022 saw just three peak-time diagrams with pairs of 319 EMUs.

With just three months left of London Midland, 350248 works the 2T69 14.34 Euston to Tring service at Bourne End on 9 September 2017. The 350/2s are different to the other 350s because they have a 2+3 seating arrangement. They were built with a top speed of 160km h/100mph but were later upgraded to 177km h/110mph running so they could work services with other 350 sub-classes. After the 730s have entered service, the entire Class 350/2 fleet will be transferred back to Porterbrook for possible conversion to battery power.

In the new LNR livery, 350118 leads 350107 working 1Y24 14.22 Northampton to Euston at South Kenton on 28 May 2022. The 30 Class 350/1s are dual voltage and a handful enjoyed a stint on hire to Southern in 2009 working East Croydon to Milton Keynes services, covering for several Class 377/2s on hire to First Capital Connect due to delays in the delivery of its Class 377/5s.

LNR outgoing-liveried 350244 leads 350111 in LNR corporate colours working 1Y26 15.25 Northampton to Euston fast service over the Grade II listed Bushey arches railway viaduct on 16 June 2022. The 350/2 fleet is still carrying either the London Midland livery (with LNR branding applied) or an outgoing variation of the LNR livery.

Now with LNR branding, 319429 in the remnants of London Midland livery leads ex-Thameslink-liveried 319214 working 2B14 07.39 Bletchley to Euston at Kings Langley on 29 July 2021. The 319s are now only utilised for peak time services.

Four LNR 319s had been taken off lease early in 2022 and sent to Newport Docks by road for scrapping after component recovery at Burton Wetmore depot. Ex-Thameslink-liveried 319005 leads ex-LM liveried 319429 working 2K29 17.23 Euston to Milton Keynes service, passing Wembley yard and crossing the Watford DC line nearing Wembley Central on 31 August 2022.

London Overground

London Overground (LO) established in 2007 by Transport for London (TfL), took over and upgraded Silverlink London metro services, stations and facilities on the North London Line, Watford DC line, Gospel Oak Barking line and West London line. The aim was to create a modernised suburban orbital railway by incorporating and building extensions to other lines with frequent services serving parts of London that had suffered neglect and a lack of transport provisions.

London Overground is a TfL service but is operated by Arriva Rail London. TfL has a current fleet of 57 five-car Class 378 Capitalstars introduced between 2009 and 2011, some originally as three-car trains and were all extended to five-cars by 2015. Thirty-seven of the fleet are dual-voltage AC/DC 387/2s (378201–234, 378255–257) and 20 are DC-only 378/1s (378135–154). These were built with metro-style longitudinal seating.

London Overground has 30 four-car AC-only 710 Aventra units (710101–130), 24 dual-voltage units divided into 18 four-car (710256–273) and six five-car (710374–379), built by Bombardier (later Alstom) at Derby Litchurch Lane, which has replaced Class 172, 315 and 317 rolling stock on the Gospel Oak to Barking line, Romford to Upminster line and Lea Valley services. They have also been introduced on the Watford DC line to free up Class 378s for other services.

The East London line, a tube line between New Cross, New Cross Gate and Whitechapel (previously Shoreditch until June 2006) was closed in December 2007, to allow extension works to connect Dalston Junction in the north by reviving the existing trackbed along Kingsland Viaduct, which was mothballed in June 1986, once carrying services to the now extinct Broad Street station.

The East London line would then be extended south via a new connection to the existing Brighton Main Line to West Croydon with branches to New Cross and Crystal Palace. The first train to test the new extension travelled on 5 October 2009. The East London line was opened to the public on 27 April 2010 with a preview weekday service between New Cross Gate and Dalston Junction, with a full-service commencing to West Croydon and Crystal Palace from 23 May 2010.

In February 2011, the line was extended to connect to the North London line from Dalston Junction to Highbury and Islington. In December 2012, a branch was built south of Surrey Quays connecting the South London line enabling trains to run through to Clapham Junction. This completion created an inner orbital line around London.

On 31 May 2015, London Overground took over the West Anglian services to Chingford, Cheshunt and Enfield Town via Seven Sisters and services between Romford and Upminster. The first 17 Class 315s (315801–817) transferred from Greater Anglia, with the remaining 44 315s (315818–861) moving to TfL Rail. The four-car 315s were built by BREL at York Holgate between 1980–81, replacing the Class 306s on GEML metro services, followed by suburban West Anglia services in 1984.

Eight four-car Class 317/7s (708–710, 14, 19, 23, 29 and 32) and six four-car Class 317/8s (887–892), which had been replaced on Stansted Express services by 379s, were transferred to London Overground alongside the 17 315s. These were all replaced on the Lea Valley routes by Class 710s from March 2020, after delays getting the new EMUs into service because of software issues.

Leased by Angel Trains, London Overground had a fleet of eight Class 172/0s (172001–172008) that worked the Gospel Oak to Barking line (known as the 'GOBLIN') that had replaced Class 150s in May 2010. As part of CP5, Network Rail awarded a £56.9m contract to J Murphy and Sons Ltd to electrify the line, although delays through incorrect overhead structures and late delivery of materials meant completion didn't happen until mid-January 2018 with 378211, the first EMU, testing the overhead wires on 11 January 2018. Delays in the arrival of new Class 710 EMUs meant the lease of 172s was extended twice. Only 172002 went off lease in July 2018, followed by another in November 2018. The transfer of a third Class 172 to WMR in late January 2019 meant London Overground no longer had enough DMUs to fulfil its 15-minute-interval timetable on the GOBLIN. With the remaining Class 172s needing to transfer to WMT after an extended stay and with software glitches further preventing the introduction of the eagerly awaited new Aventra 710s. TfL and Arriva Rail London opted to shorten three five-car Class 378s (378206, 378209 and 378232) to four-car formation (to fit the platforms) and run a reduced 30-minute interval timetable from mid-March 2019 as a stop-gap measure. 378232 was the first EMU to enter service on the GOBLIN on 28 January 2019, a year after the line was electrified. After months of delays and overcrowding, TfL offered one month's free travel for passengers from 31 August to 1 October 2019 as an apology, shortly after normal timetabling had been restored with eight 710s in service.

The line between Richmond and Gunnersbury is shared with London Underground's District Line as identified by the fourth-rail system. Trains between Richmond and Acton Central run on 750 V DC third-rail supply before switching to 25 Kv AC overhead supply through to Stratford, therefore trains must be dual-voltage. With the Strand-on-the-Green as the backdrop, 378213 powers 2N07 17.40 Stratford to Richmond passing 378205 working 2N58 18.32 Richmond to Stratford on the Grade II listed Kew Railway Bridge on 17 June 2022. This is one of only three structures that crosses the Thames on the London Overground network alongside the Rotherhithe tunnels and Battersea Railway bridge.

Crossing the ECML on the four-track section of the North London line at Caledonian Road & Barnsbury, 378229 is working 2N29 08.40 Willesden Junction to Canonbury on 7 February 2020. Many freight trains also work on this section.

Framed between the Nike factory store and Hackney Church, 378214 works 2N23 10.47 Richmond to Stratford to the east of Hackney Central on 18 May 2022. Services would previously continue beyond Stratford to North Woolwich, but this was withdrawn on 9 December 2006 as the line was duplicated by the Jubilee Line and Docklands Light Railway.

Working on the relatively short West London line section between Willesden Junction and Clapham Junction, 378256 powers the 2Y76 10.21 Stratford to Clapham Junction service over Battersea Railway bridge on 2 November 2021. The train has just left Imperial Wharf station, which was opened on 27 September 2009 to serve the regenerated area of the former gas works at Chelsea Harbour, now an urban-riverside village.

Several 378s have received a general refresh at Ilford, wearing the new London Overground colour scheme, as carried by the Class 710s. 378135 working 9H10 07.45 Clapham Junction to Dalston Junction service, approaches Wandsworth Road at Factory Junction on 10 August 2022. The two lines to the left go to Battersea Park station, which has a very limited overground service. During engineering works or problems at Clapham Junction, overground services can terminate at Battersea Park instead.

Working on the South London line, 378153 powers the 9H27 11.59 Clapham Junction to Dalston Junction service over Brixton station on the 'Atlantic flyover' on 24 April 2022. The Overground network passes over both Brixton and Loughborough Junction stations but due to the tracks elevated on a curved viaduct with a gradient, the construction and associated costs of high-level platforms are not feasible.

As part of the East London line extension to connect with the former North London line branch to Broad Street, a new bridge was constructed to carry the lines over the GEML. Appearing out of the boxed Shoreditch High Street station, 378218 is working 9G36 14.19 Dalston Junction to Clapham Junction on 5 July 2022. Shoreditch High Street station was built on the site of Bishopsgate goods yard, which was destroyed by fire in 1964. It replaced the original Shoreditch station to the east and was built in a boxed tunnel so future development works in Bishopsgate can take place without the need to close the railway.

378141 descends the ramp that takes the line from the bridge over the GEML south of Shoreditch High Street to the sub-surface tunnels at Whitechapel. The DC-only Capitalstar is working 9G31 13.04 Dalston Junction to Clapham Junction, passing the graffiti artists at Brick Lane Station Park near Whitechapel on 11 July 2022. The new line is to the right of the now filled-in Shoreditch station of the East London line, which was closed in June 2006.

315806, with its twin high-intensity headlights, works 2U49 14.22 Enfield Town to Liverpool Street approaching Rectory Road on 26 June 2018. 315804, 806, 809 and 812 all had modified headlights as part of a trial to increase the front-end light visibility of 1972-designed EMUs. The London Overground 315 fleet was retired from service on 20 October 2020 having been displaced by new Aventras 710/1s. 315806 was taken to Newport Docks for scrapping on 31 July 2020.

317710 leads 317891 working 2U57 15.52 Enfield Town to Liverpool Street at Cambridge Heath on 15 September 2019. By April 2020, the Overground 317s had been displaced by new Class 710s. Delays in the introduction of new 720s prompted Greater Anglia to reinstate eight of the former Overground 317/7s to service until 31 May 2021. These had been in warm storage at Wembley Yard and were chosen because they were considered more PRM-compliant than some of Greater Anglia's own 317s, which had not been fitted with universal accessible toilets. All Overground 317s have since been scrapped.

172007 works 2J33 10.20 Gospel Oak to Barking at Walthamstow Wetlands near Blackhorse Road on 5 March 2019. The two-car Diesel Turbostars were often overcrowded at peak times and all eight were due to be transferred to West Midlands Trains (WMT) in June 2018, however, the last three didn't transfer until mid-March 2019. They were replaced temporarily by three four-car 378s before the arrival of the delayed Aventras.

378206 works 2J73 15.20 Gospel Oak to Barking into Wanstead Park on 24 May 2019. Introduced in late January 2019, the Capitalstars remained on the 'GOBLIN', running a reduced 30-minute interval timetable from mid-March 2019 following the cascade of 172s to WMT. The Aventras finally made their debut on the GOBLIN on 23 May 2019, and by August 2019, all eight 710s were in service and normal timetable was restored.

There are 18 four-car, dual-voltage Bombardier Class 710/2 with eight required for use on the GOBLIN. The Gospel Oak to Barking line was later extended to Barking Riverside with trains sharing the London, Tilbury and Southend (LTS) line through platform 7/8 at Barking before diverging at Ripple Lane onto a new 1.6km/1-mile branch line situated on a viaduct to the elevated terminus at Barking Riverside. Barking Riverside is the 113th London Overground station, opening on 18 July 2022 and will serve a new town, regenerating the area previously occupied by Barking Power Station with 10,000 new homes. 710273 works 2J24, the 08.36 Barking Riverside to Gospel Oak service, shortly after departing Barking Riverside at Northgate on 31 August 2022.

The 5km/3-mile rural Romford to Upminster single branch line was taken over by London Overground (LO) from 31 May 2015 despite being separated from the rest of the LO network. The branch has two services per hour in each direction, scheduled to take nine minutes and is restricted to 30mph running. 710108 nears Emerson Park working 2V32 13.41 Romford to Upminster on 11 June 2022. A new Class 710 Aventra operated the diagrams from 5 October 2020, replacing the previous Class 315 or 317.

New Class 710/1s entered service on the Lea Valley routes, formally on 3 March 2020 after a brief debut on 24 February 2020, replacing ageing 315 and 317 units. Operating in a pair as an eight-car service, 710259 leads 710117 over the River Lee Navigation (a canalised river) approaching Clapton on 5 July 2022. The 710/1s and 710/2s can be used interchangeably on the Romford/Upminster branch, Lea Valley and GOBLIN services although 710/2s are preferred on the latter as they can access Willesden Traction Maintenance Depot via Willesden low level using DC pickup.

Class 710s made their debut on the Watford DC line on 9 September 2019. All 710s operating on this route are either four-car 710/2s (710256–273) or five-car 710/3s (710374–379) because of their dual-voltage. The introduction of Aventras on the Watford DC line has enabled 378s to be cascaded elsewhere. 710265 powers 2C21 10.30 Watford Junction to Euston at South Hampstead on 16 September 2022. This is one of two places where the Watford DC line passes below the Chiltern Main Line.

Railtours

This chapter showcases some of the special excursion trains that operate in London. Locomotive Services Ltd and West Coast Railways have bases at Southall and operate several steam and diesels tours out of the capital each year. The luxurious Belmond British Pullman is serviced and maintained at Stewarts Lane, Battersea and is worked by DB Cargo Class 67s or steam (usually Clan Line) to destinations mainly around the southern and south-eastern region of England. This includes people travelling the Venice Simplon-Orient Express between London and Folkestone for onward coach travel via the Channel Tunnel shuttle. The Hastings DEMU (diesel-electric multiple unit) is based at St Leonards so often makes excursions through the capital. There have also been classic Buffer-Puffer railtours operated by Pathfinder Railtours, which take in unusual track around London. UK Railtours, based in Hertfordshire, often commences services from Finsbury Park or Kings Cross, usually to destinations in the north of England. More recently, Locomotive Services Ltd (LSL) has operated InterCity charters out of London Euston and has acquired The Steam Dreams Rail Co and expects to provide its future Royal Windsor Steam Express trains.

The classic Hastings DEMU, classified on the Total Operations Processing System (TOPS, the computer system for managing locomotives) as 201001, incorporates vehicles from several units that worked for British Rail between London and Hastings between 1957 and 1986. On 3 September 2022, DEMU 1001 powers the 1Z23 07.45 Tonbridge to Portsmouth Harbour via Kensington Olympia and Waterloo on 'The Vectis Venturer' UK railtour, departing Vauxhall on the South West Main Line.

Locomotive Services Ltd (LSL) have impeccably revived the iconic 1960s Midland Pullman using an InterCity 125 HST. Here 43047 leads 43046 working 1Z44 07.01 Eastleigh to Scarborough *Yorkshire Coast and Jorvik Pullman* on the North London line at Willesden Junction on 27 August 2022. The luxurious train is passing over the WCML.

The Class 50 Alliance Ltd, 50007 *Hercules* in GB Railfreight colours, powers over the Thames at Grosvenor Bridge working 1Z50 08.00 Victoria to Exeter St Davids *Devon Pullman* on 18 June 2022. The Belmond British Pullman, established in 1982, offers luxurious rail travel around the UK and takes passengers to Folkestone as part of the Venice Simplon-Orient Express using DB Class 67s or Clan Line. It is based at Stewarts Lane, Battersea with trains working out of Victoria. On this occasion, the stock was being used by UK Railtours.

Many steam tours operate out of Victoria, operated by West Coast Railways (WCRC), DB Cargo and LSL. The trains are usually brought in ECS by a diesel as there is no run-round. WCRC 33207 *Jim Martin*, which is no stranger to the southern region, brings in 5Z66 05.38 Southall to Victoria ECS over Grosvenor Bridge on 13 September 2018. A4 60009 *Union of South Africa* is on the rear and would work the train to Swanage. The A3216 Chelsea Bridge and B304 Albert Bridge can be seen in the distance.

With pathing difficulty and Victoria being better equipped for rail-excursion preparation, not many rail tours originate from Waterloo. On 15 July 2022, UK Railtours' 'The Solent Searcher' tour started from Waterloo taking in rare branch lines at Marchwood, Hampshire and Ludgershall Ministry of Defence (MoD) Wiltshire. GBRf 69004 leads 66799 away from Waterloo working 1Z30 08.14 to Marchwood.

The 'Royal Windsor Steam Express' was launched by Steam Dreams in the summer of 2019, with one-way trips to Windsor and an evening dining train often doing a circular trip via the Chertsey loop. In its first year, trains originated out of Waterloo mainly using LNER Thompson Class B1 61306 *Mayflower*, top-and-tailed with a WCRC Class 33 or 47. Trips have been affected by the Covid-19 pandemic and a change of operator in 2022, but it is hoped these trips will return in 2023 with LSL. Future trips are booked out of Victoria. With 33207 on the rear, 61306 *Mayflower* pulls away from Waterloo with 1Z80 08.05 to Windsor & Eton Riverside on 6 August 2019.

Southeastern

Southeastern (SE) operate main line and metro services between London, Kent and East Sussex including domestic high-speed services on HS1 under the Integrated Kent franchise. Connex Southeastern operated the franchise from its inception on 13 October 1996, inheriting rolling stock and services between London and Kent from NSE. Connex Southeastern was stripped by the Strategic Rail Authority (SRA) of the franchise following poor financial management, ceasing operations on 8 November 2003. The franchise was brought into public ownership on an interim basis, ran by South Eastern trains on behalf of the SRA. This was until London and South Eastern Railway Ltd owned by Govia (a joint venture between Go Ahead and Keolis) won the franchise and commenced operations from 1 April 2006 under the name Southeastern.

After several contract extensions, the franchise was terminated early on 16 October 2021 following a breach of trust with the DfT after failing to declare more than £25m of taxpayer's funding. The franchise was returned to public ownership as SE Trains Ltd, a subsidiary of the DfT, trading as the same name Southeastern.

To replace the ageing 4EPB slam-door stock, a fleet of 147 four-car Class 465 Networkers were built between 1991 and 1994 originally for NSE. The order was split between two manufacturers with the 465/0s (465001–050) and 465/1s (465151–197) being built by BREL/ABB at York, Holgate and the 465/2s (465201–250) built by Metro-Cammell at Birmingham, Washwood Heath. In total, 43 two-car Class 466s (466001–043) were also built by the latter to work either with 465s in ten-car formations or solely on branch-line services. Due to non-compliance with accessibility standards from January 2021, 466s can now only work in service coupled to a 465. The 465/0s and 465/1s were upgraded with Hitachi AC traction equipment to improve reliability at Ashford between 2009 and 2010, followed by a mid-life refurbishment programme carried out by Railcare.

In 2005, 465201–234 underwent extensive refurbishment at Wabtec Rail Doncaster with 12 first-class seats fitted in each driving car. These were renumbered as 465/9s (465901–934) and were deployed on SE outer-suburban trains. The remaining 465/2s have since been stored at Worksop, later Ely, alongside a handful of 466s with an uncertain future.

Bombardier, Derby built 36 five-car Class 376s (376001–036) between 2004 and 2005 to work on inner-urban metro services. These spacious EMUs were the last in the UK to be delivered without air-conditioning. They have no toilets or plug doors and are due for refurbishment.

The main traction for South Eastern Main Line services are Class 375 Electrostars with 112 built between 1999 and 2004 by Adtranz/Bombardier Derby Litchurch Lane to replace post-privatisation slam-door EMUs. Southeastern operate ten three-car 375/3s (375301–310), 30 four-car 375/6s (375601–630), 15 four-car 375/7s (375701–715), 30 four-car 375/8s (375801–830) and 27 four-car 375/9s (375901-927) with slight variations among the five sub-classes. The entire 375 fleet was refurbished at Derby Litchurch Lane between 2015 and 2018 including a repaint into the SE dark blue livery

Built for Southern between 2008 and 2009 at Bombardier Derby, 23 four-car Class 377/5 (377501–523) were transferred to SE between 2016 and 2017 after being leased to FCC, later TL. They were joined by Southern 377163 and 377164, although these returned to GTR after the summer timetable change in 2022. The 377s mainly work on Maidstone-line services to Victoria and Blackfriars. It is believed some may return to GTR when more 707s are cascaded from South Western Railway (SWR).

SE operate the UK's first domestic high-speed service using 29 six-car dual voltage Class 395s (395001–029), built by Hitachi at Kasado, Japan, which entered service from June 2009. They operate on HS1 reaching speeds of 140mph and work on the classic main line to Dover and Ramsgate. These were the first Hitachi units to operate in the UK and are nicknamed 'Javelins' because they were part of the transport infrastructure upgrade programme for the 2012 Olympic Games at Stratford.

The 29 Class 395s will go through a £27m upgrade programme from March 2023 including installation of new seats, LED lighting, USB ports and live passenger information systems.

Built by Siemens, Krefeld in Germany, the 30 five-car Class 707s (707001-030) at SWR will all transfer to SE for use on its metro services between Dartford, Sevenoaks and Hayes. The Desiro City EMUs named CityBeam entered service with Southeastern on 27 September 2021 in corporate colours. 707014–24 and 707030 are yet to transfer due to delays in introducing the new 701s into service.

The oldest members of the SE fleet are the stalwart 1991–94 built Class 465/466 Networkers. 465171 leads 465158 working 2S38 13.16 Charing Cross to Sevenoaks approaching the Bermondsey dive-under on 7 August 2022.

Working out of Blackfriars because of planned engineering works at Victoria, 465169 leads 465184 working 2M47 15.56 Blackfriars to Orpington approaching the former Walworth Road station (closed in 1916) south of Elephant and Castle on 7 August 2022. Although services on the Holborn Viaduct to Herne Hill line are predominately Thameslink trains, Southeastern runs peak services out of Blackfriars to Ashford and intermittent trains to Maidstone East. From 11 December 2022, a new timetable will see services between Maidstone East, Ashford and Blackfriars removed and replaced with just a few peak-time services to Beckenham Junction and Dartford.

466022 leads 466035 and 456002 over Cannon Street railway bridge working 2I55 17.38 Cannon Street circular service, taking the Sidcup line in an anti-clockwise direction and returning via the North Kent line on 14 September 2022. Built on the site of a medieval steel yard, Cannon Street is situated off a triangular junction between London Bridge and Charing Cross and is identified by its two Grade-II listed Wren-style towers, which once supported an iron train shed that was demolished in 1958. This was followed by the Southern Station House, which fronted the station in 1963. The towers and part of the side walls of the station building are all that remain of the original architecture since opening on 1 September 1866. The station has gone through major redevelopment with construction of multi-storey office buildings throughout the 1960s and 1980s.

A trio of Go-Ahead trains passing Grosvenor Road Carriage Sheds on 8 June 2021. Southeastern 465173 and 465177 are working 2M20 07.58 Victoria to Orpington. Southern 377614 and 377621 are working 1I07 06.45 Horsham to Victoria, and Gatwick Express 387209 and 387213 are working 1W22 07.59 Victoria to Brighton 'Southern' service. This is the site of the Grosvenor Road Station, which closed in 1911.

376009 and 376010 work 1V28 12.37 Charing Cross to Hayes, Kent at Clock House on 13 May 2022. The line from Courthill Loop Junction North, Lewisham to Hayes is called the Mid-Kent line or just the Hayes line. Despite no longer being in Kent, Hayes is referred to as Hayes (Kent) by National Rail timetables to avoid confusion with Hayes and Harlington on the Elizabeth line.

Shortly after crossing Hungerford Bridge, 376017 leads 376035 working 2N66 12.48 Charing Cross to Gravesend, passing the BFI IMAX cinema on the approach to Waterloo East on 9 August 2022.

376004 leads 376008 with 2L30 11.12 Charing Cross to Dartford into Waterloo East station on 9 August 2022. This is taken from the footbridge, which leads to the 1990 circa high-level bridge that provides pedestrian access between Waterloo East and Waterloo Main. The tubular walkway is elevated above the alignment of the old railway linking the London South Western Railway and South Eastern Railway. The line, constructed in 1864, emerged from around where Platform 10 is now at Waterloo, extending across the concourse and station frontage connecting to the SER before it was discontinued during station rebuilding in the early 20th century.

Passing Battersea Dogs & Cats Home, 375821 leads 375814 working 1Y46 15.08 Victoria to Ramsgate on 10 September 2022. This is the location of Battersea Park Road Station, which closed in 1916.

With the London Shard dominating the background, 375815 leads 375810 working 1H14 07.15 Charing Cross to Hastings 'main line' service, and 376001 leading 376008 working 2S15 07.20 Cannon Street to Orpington 'metro' service away from London Bridge on 5 July 2022.

The 23 SE 377/5s, which transferred from GTR, may often be found on the northern end of the Brighton Main Line, running empties to be maintained and serviced by Southern at Selhurst depot. 377521 is working 5Z50 09.06 Selhurst to Stewarts Lane Depot on the non-passenger stretch between Pouparts Junction and Longhedge Junction, nearing its destination on 7 August 2022. Southern crew would later take the EMU to Victoria and hand it back to Southeastern.

Javelin 395022 *Alistair Brownlee* powers 1J03 05.13 Ashford International to St Pancras International between Thames Tunnel and Purfleet on 28 June 2022. The viaduct passes over the Dartford Tunnel exit roads and the LTS line and under the iconic Queen Elizabeth II Bridge.

707007 leads 707004 working 2O37 11.35 Cannon Street circular via Greenwich and returning via the Sidcup line approaching Greenwich Station on 6 October 2022. The CityBeam units are crossing the now disused 1963 lifting bridge, which would raise a section of track for tall boats navigating Deptford Creek, the northernmost part of the River Ravensbourne. The structure is illustrious of Greenwich's marine and industrial past. The line between London Bridge and Greenwich is believed to be the world's first suburban railway, opened in phases between April 1834 and December 1838 and is entirely elevated on a bricked viaduct.

Southern and Gatwick Express

Govia Thameslink Railway (GTR) a Go-Ahead company, operates the Thameslink, Southern and Great Northern franchise (TSGN), with services on the Southern part of the franchise operating under the trade name 'Southern'. Southern has been operating the South Central franchise since it took over from Connex South Central on 26 August 2001, having bought the remainder of Connex's franchise, which was due to be terminated in May 2003. Southern was later absorbed into the TSGN franchise, which Govia retained from 14 September 2014.

Part of the original franchise commitment was to replace the ageing MK1 slam-door Class 421 and 423 rolling-stock, which were withdrawn in 2005. The core of Southern's fleet became the ubiquitous Class 377 Electrostar, built by Bombardier at Derby Litchurch Lane and can be found anywhere on the Southern network except on the non-electrified Uckfield branch and Marshlink line.

The first to be the built in 2001–02 were 28 three-car DC-only 375/3 (375311-338), which were later reclassified as 377/3 (377301-328) following the fitting of Dellner couplings. These initially worked on Southern Coastway lines but were replaced by cascaded 313s from London Overground from 2010 so they could strengthen London suburban services.

Closely following were 64 four-car DC-only 377/1s (377101–164), 15 four-car Dual-voltage 377/2s (377201–215) and 75 four-car DC 377/4s (377401–475), built between 2002 and 2005. In addition, 23 five-car Class 377/5s (501–523) were built between 2008 and 2009, intended for Southern to replace its First Capital Connect (FCC)-bound Class 319 fleet but were instead also subleased to FCC. Due to delays in finalising new trains for Thameslink, Southern procured more 377s. There was an additional order of 26 DC five-car 377/6s (377601–626) and 8 dual-voltage five-car 377/7s (377701–708) between 2012 and 2014, which would lead to the cascade of the 24 two-car 456s to SWR. With 377163 and 377164 returning from Southeastern in May 2022, Southern now operate a total of 216 377s.

Southern also inherited 46 four-car Class 455s (455801–846) from Connex used predominately on suburban and metro services. Built between 1982 and 1984 at BREL Holgate works, York these DC MK3 workhorses were retired from service on 14 May 2022 prior to the summer timetable change, with 377s replacing them. All have since been scrapped at Newport Docks, with 455838/839 the first pair to leave Stewarts Lane Depot on 4 May and the last pair 455819/804 on 16 August.

Southern has a fleet of Bombardier-built Class 171 Turbostars for its non-electrified lines. These were built at Derby Litchurch Lane between 2003 and 2004. They are identical to a Class 170 apart from the Dellner coupling, which can be used to attach to 377s in an emergency. The two-car 171/7s and four-car 171/8s were built for Southern between 2003 and 2004. The two-car 171/2s (201 and 202) and four-car 171/4s (401 and 402) were formed from four cascaded Scotrail three-car 170s between 2016 and 2018. Towards the end of 2022, The Southern Class 171 fleet was reshuffled. The former Scotrail 171s reverted to three-car 170s and sent to EMR except for 171201, which is remaining on sub-lease. Six of the two-car 171/7s and all six four-car 171/8s will be reformed to three-car, creating a fleet of 13 three-car 171s (inc. 171201) and four two-car 171s to help operational flexibility.

The only DMUs in the Govia Thameslink Railway (GTR) fleet are the Class 171s, which work London Bridge to Uckfield services having replaced the Class 205 and 207 Thumpers in 2004. They also work on the Marshlink line. GTR's six four-car Turbostar fleet will be reformed to three-cars in late 2022, with the spare middle carriage being used to reform six two-car units to three-car units. Soon to be reformed as a three-car set, 171802 working in four-car formation powers the 1E29 12.08 London Bridge to Uckfield over Riddlesdown Viaduct on 6 November 2017.

455807 leads 455834 working 2B31 09.08 Epsom Downs to Balham approaching West Croydon and passing the turnback siding (the middle track) on 30 April 2022. Engineering works taking place outside Victoria meant trains were terminating at Balham. Notice the removal of cab end gangways with the space used for driver's cab air-conditioning.

455805 leads 455815 working 2T09 08.28 East Croydon to London Bridge service passing Selhurst Train and Rolling Stock Maintenance Depot on the approach to Norwood Junction on 12 May 2022. GTR's Selhurst Depot is one of the busiest in the UK, with several high-tech facilities. It lies in the triangular area between Selhurst and Norwood Junction stations, just north of East Croydon. Behind the building on the right, are the North sidings and adjacent cleaning shed and repair shop, which are located on the site of the former Croydon Common Athletic football ground where Crystal Palace played matches between 1918 and 1924.

455833 leads 455815 off the spur from Tulse Hill working 2H37 14.31 London Bridge to Beckenham Junction at West Norwood Junction on 30 April 2022. This service was cancelled at Birbeck due to late running with passengers using the tram to continue their journey. The lines to the left continue to Streatham Hill.

455833 leads 455822 working 2B18 19.00 Victoria to Epsom Downs, arriving into Clapham Junction on 12 May 2022. The train is passing the Clapham Junction 'B' Signal Box, which opened on 12 October 1952 and ceased signalling trains in November 1980 when the signalling for the area was moved to Victoria Area Signalling Centre. The box is now used as an office for Network Rail operations and maintenance staff. Under Phase 3 of the Victoria Re-signalling Programme, set to be completed in December 2022, the signalling and control for the area covering Victoria terminus, Clapham Junction (central side) and Balham will transfer to the rail operating centre (ROC) in Three Bridges with a complete renewal of the 1980s signalling equipment, including replacing the track circuits with Frauscher axle counters. Work on Phase 3 started in early 2021 and includes 93 signals being converted to LEDs; 17 signal gantries and 50 points being renewed, including the ladder between the fast and slow Brighton lines north of the station being remodelled to increase line speeds on the crossovers.

On 14 May 2022, GTR in partnership with the Branch Line Society, ran a charity farewell tour named the Metro Marauder to commemorate the last day of Southern 455s after 39 years in service. The tour started at Victoria visiting Brighton via the Mole Valley, Charing Cross, Blackfriars, and Sevenoaks as well as taking some unusual track before returning to Victoria. 455835 and 455841 curve out of Waterloo East working the 1Z57 12.55 Brighton to Charing Cross leg. The sell-out tour was greeted by many enthusiasts and raised more than £26,000 for Mind Croydon, a mental health charity.

Left: South Bermondsey is the nearest station to The Den football stadium for Millwall supporters travelling to their home games. In familiar surroundings, 455841 leads 455835 working 1Z58 15.31 Charing Cross to Blackfriars Metro Marauder farewell tour passing The Den on 14 May 2022. The last Southern 455 passenger service was in the early hours of 15 May 2022 with 455843 working 2U76 00.25 Victoria to Norwood junction via Crystal Palace. All the SN 455s have since been scrapped at Newport Docks.

Below: In April 2020, GTR applied NHS 'We Thank You/We're With You' branding to 377111, 700111 and 717017 to showcase its support for NHS key workers during the Covid-19 pandemic. NHS-branded 377111 leads 377472 working 2B39 10.08 Epsom Downs to Victoria at Pouparts Junction, Battersea on 29 July 2022.

Almost simultaneous departures from Clapham Junction, as 377148 leads 377150 on 2I11 08.31 Epsom to Victoria service and 377457 leading 377436 and 377412 working 1C13 07.29 Bognor Regis and Southampton Central to Victoria service on 7 May 2022. The Brighton Main Line is one of the most congested railway routes in the UK, with main line and suburban services out of London Bridge and Victoria serving Gatwick Airport, Brighton and linking the Southern home counties.

Five-car 377707 working 2J51 17.44 London Bridge to Caterham exits the ornate southern portal of Knights Hill Tunnel as it arrives at Tulse Hill on 9 July 2022. Knights Hill Tunnel is a 303m/331-yard tunnel between North Dulwich and Tulse Hill, which was opened in 1868 by the London, Brighton and South Coast Railway. The lines that diverge to the left are the Thameslink lines towards Herne Hill.

Following the withdrawal of the 455s and the 2022 summer timetable change, Milton Keynes services were cut back to Watford Junction with limited peak services to Hemel Hempstead and often curtailed to four-car trains. 377205 passes under the Circle and Hammersmith & City lines on the approach to Shepherds Bush with 2O49 15.52 Watford Junction to East Croydon on 7 October 2022. Only dual voltage 377s (377/2 or 377/7) can work West London line services, with the switch-over from DC third rail to AC overhead pick-up taking place at North Pole Signal VC813 going north and VC818 going south.

Above: St George Wharf Tower at the time of completion in 2014, was the tallest residential building in London and can be seen in the left side of the photograph alongside the new high-rise buildings in Vauxhall and Nine Elms. Chelsea Road Bridge offers spectacular views of the River Thames and the regenerated former industrial area around Battersea Power Station as 377603 works the 2I58 20.11 Victoria to Epsom across Grosvenor Bridge on 9 July 2022.

Left: With the heavily regenerated Battersea Power Station area in the background, 377626 and 377707 form a ten-car service working 2B15 18.41 Sutton to Victoria across Grosvenor Bridge on 7 August 2022. The 377s took over all metro services with the retirement of the 455s from the 2022 summer timetable change.

With the SELCHP incinerator in the background, Southern 377624 works 2F39 15.28 London Bridge to Crystal Palace service into New Cross Gate on 7 August 2022. The Electrostar has just gone under the New Cross Gate flyover, which allows Overground trains from West Croydon and Crystal Palace to connect to the East London line (ELL) towards Surrey Quays without conflicting the main line. New Cross Gate depot, opened in 2010, is the main depot for stabling and servicing Class 378s on the ELL.

Gatwick Express

The initial standalone Gatwick Express (GX) franchise was incorporated into the South-Central franchise operated by Southern in June 2008, with services being extended to Brighton from December 2008. Southern leased 17 Class 442s, previously mothballed by South West Trains, for the new services with each receiving an internal makeover. When Southern retained the franchise in June 2009, the remaining seven 442s were also refurbished and returned to service leading to the gradual withdrawal of the 460 Junipers by September 2012. The Wessex Electric EMUs continued until they were replaced by new Class 387s in 2016. 442413 worked the final booked 5-WES Gatwick Express services on 17 September 2016.

Despite finishing on the GX services in September 2016, Southern retained six Class 442s for use on four peak-time diagrams to Brighton and Eastbourne in and out of London Bridge until March 2017. Built by Bombardier at Derby Litchurch Lane, 27 four-car Gatwick Express Class 387/2s (387201–227) gradually replaced the Class 442s on Gatwick Express services throughout 2016. 387211 and 387207 were the first to enter service with GX on 29 February 2016, although several were on hire to Thameslink a few weeks prior due to delays in commissioning the 700s. Gatwick Express services were suspended in March 2020 and resumed briefly in December 2021 before the Omicron Covid variant led to another suspension. Services finally resumed on 3 April 2022, albeit with only two trains per hour (instead of four) due to the Gatwick Airport Station upgrade.

On 19 May 2016, GX 442419 works a Gatwick Express service into Victoria. Notice the removal of the 'Gatwick' branding from the bodyside to not confuse passengers using the trains on the southern end of the Brighton Main Line.

442410 and 442413 await departure from London Bridge with 1B47 17.57 service to Brighton – the last Southern Class 442s in revenue-earning service on 10 March 2017. All the 5-WES EMUs were sent to storage at Ely Papworth sidings before making a remarkable but sadly short-lived comeback with South Western Railway.

Gatwick Express 387227 leads 387215 working 1W72 13.59 Victoria to Gatwick Airport at Pouparts Junction, Battersea on 27 August 2022. The Express is passing over the West London lines and the lines from Waterloo can be seen in the background to the left.

Under a threatening sky, 387219 leads 387226 and 387212 working 1W14 17.59 Victoria to Brighton over Grosvenor Bridge shortly after leaving Victoria on 10 September 2022. It would take just half an hour for the train to get from Victoria to Gatwick.

South Western Railway

South Western Railway (SWR), a consortium of First Group (70 percent) and Mass Transit Railway (MTR) (30 percent) commenced operation of the South Western franchise on 20 August 2017. Its predecessor, South West Trains (SWT), owned by Stagecoach, ran the franchise from 4 February 1996. SWR is contracted to run the South Western franchise until at least May 2023. It operates the intense suburban and metro network around south-west London and into the southern home counties. As part of the franchise commitments, SWR ordered 60 ten-car and 30 five-car Class 701s from Bombardier (later Alstom), financed by Rock Rail to replace its Class 455, 456, 458 and 707 fleet inherited from SWT on suburban and metro services.

With dwindling passenger numbers because of the Covid-19 pandemic, SWR released an amended timetable on 17 January 2022, which saw the retirement of its 1990–91-built two-car Class 456 (456001-024) fleet two days prior. All Class 456s have since been stored and subsequently scrapped, with 456003 the first to leave for Long Marston on January 18, 2022.

SWR originally had 91 Class 455 sets (which were ordered to replace the 4-SUB/EPB units in the mid-1980s), but several have already been withdrawn and scrapped. The Class 455/8s were built by BREL at York between 1982 and 1984, with 46 (455801–846) for Southern and 28 (455847–874) for the South Western division, with a follow-on order of 43 455/7s (455701–743, with 455743 later being renumbered to 455750) built between 1984 and 1985 and 20 455/9s (455901–920) built in 1985. The 455/7s were produced as three-car sets and received a trailer standard open (TSO) taken from Merseyrail-bound Class 508s to form four-car sets. The ex-508 trailers can easily be identified by their body shape. The 455/9s are similar to 455/7s but have convection heating rather than pressure heating, thus have slightly different roof vents. The 455/9s were built with TSOs therefore have matching vehicle profiles, except for 455912 and 913, which have a vehicle from Class 210 stock due to accident damage.

In total, 30 Class 458 (458001–030) Junipers were built by Alstom at Washwood Heath between 1998 and 2002 to enhance capacity, although they were plagued by poor reliability. They were reintroduced in 2006 with better performance capability enabling the new Desiros to be cascaded to main-line services, allowing the first withdrawal of Class 442s in January 2007. Between 2013 and 2016, in response to rising passenger numbers, the eight eight-car ex-Gatwick Express Class 460s were combined with the mechanically similar Class 458s at Doncaster Wabtec to form a fleet of 36 five-car 458s (180 vehicles), numbered 458501–536. The four remaining 460 vehicles were scrapped. Although the Junipers will be replaced by Class 701s on metro services, SWR announced in early 2021 that it will retain 28 Class 458s (458501–528) and convert them back to four-car sets for 100mph running on the Portsmouth Direct line services instead of Class 442s, with the ability to also run them in 12 car formations.

Thirty new five-car Class 707 Desiro City EMUs (707001–030) built at Krefeld, Germany were introduced in the last stages of SWT in August 2017. All were in service by March 2018, though it was already announced these would be replaced by cheaper-to-lease Class 701s. 707001-013 and 707025-029 have been transferred to Southeastern with the rest to follow pending the introduction of 701s.

Introduced between 2002 and 2007, the most numerous fleet at SWR is the 127 Siemens four-car Class 450 Desiro (450001-127) built at Krefeld, to replace slam-door stock on outer-suburban services. From 2007, SWT renumbered 450043–070 to 450543–570 at Bournemouth with changes to the interior layout to enhance capacity, including removal of first class, for use on the Windsor lines. This was later reversed during refurbishment with SWR, with the 450/5s returning to the 450/0 series. These EMUs work on long-distance regional and suburban services within and outside of London.

Additionally, 45 five-car Class 444 Desiros (444001–045) were built by Siemens Transportation Systems in Austria between 2002 and 2004, replacing slam-door stock and Class 442 on long-distance express services between Waterloo, Weymouth and Portsmouth. For non-electrified routes to the West of England, SWR inherited 30 three-car Class 159 (159001–022 and 159101–108) and ten two-car Class 158 Sprinter units (158880–890) although 158889 is on loan at EMR. These often work in multiple together and there are currently no plans to replace them.

The 159/0s were built at BREL (British Rail Engineering Ltd) Derby Litchurch Lane between 1992 and 1993 for the NSE route to Exeter. The Class 158s were built by BREL Derby between 1989 and 1991 and eight 158/8s from TransPennine were rebuilt as Class 159/1s at Wabtec, Doncaster between 2006 and 2007. BREL Derby Litchurch Lane built 24 five-car Class 442s between 1987 and 1989 to work NSE express services between Waterloo and Weymouth. These were later replaced by new Desiro units, finishing with SWT in early 2007 and later transferring to Gatwick Express.

Upon winning the South Western franchise in March 2017, SWR had intended to return 18 of the 24 former Gatwick Express 442s to service from December 2018 to provide additional capacity on the Portsmouth Direct line. However, their re-introduction into service was over-budget, slow and staggered, plagued by safety issues with door locks, resulting in the first pair 442020 and 442010 not entering service until 10 June 2019. Further issues with electromagnetic interfere issues with lineside signals sidelined the 442s between September 2019 and January 2020.

The rejuvenation of the 5-WES units was short-lived as the fleet was withdrawn from service in March 2020 following the start of the Covid-19 pandemic and despite £45m investment, in which almost the entire fleet was fitted with new AC traction motors. In April 2021, SWR took the decision to abandon the project citing decreasing passenger numbers and the need for the 442s to receive further modifications to meet accessibility requirements from 2024. Nearly all the Class 442 vehicles have been scrapped. One driving trailer from 442401 has been preserved.

SWR operate a total of 30 Class 159s with eight of the fleet being cascaded 158s from TransPennine, which were converted and reclassified as Class 159/1 (159101–108) at Doncaster Wabtec between 2006 and 2007. All the 159/1s are still in SWR-branded Stagecoach colours. Approaching Wandsworth Town station 159107 (ex-158811) leads 159018 working 1L20 07.27 Salisbury to Waterloo on 30 January 2022. Weekend engineering works around Surbiton meant services were diverted via Richmond and Chertsey.

With Big Ben and the Houses of Parliament now dwarfed by new modern high-rise apartments dominating the skyline at Vauxhall, 159019 leads 159008 and 159012 working 1L21 09.20 Waterloo to Yeovil Junction service on 28 September 2022. All Class 159/0s are in the new SWR colours and have more powerful engines than a Class 159/1 or 158.

In the early hours of 7 August 2022, SWR 455917 and 455871, 455903 and 455912, and 455848 and 455706 are lined abreast awaiting duties at Clapham yard. The front profile of the earlier 455/8s (as for the 317/1s) includes the warning horns mounted onto the cab roof in contrast to the later 455/7s and 455/9s (and 317/2s), which have a more rounded profile with the warning horns next to the coupler.

As part of the London Waterloo upgrade in August 2017, platforms 1–4 were extended to accommodate ten-car trains resulting in many suburban services using a two-car 456 attached to two four-car 455s until the 456s were withdrawn in January 2022. Passing Wimbledon Traincare Depot on 11 August 2021 are 456022 trailing 455852 and 455729 on 2J26 10.24 Hampton Court to London Waterloo and 450116 trailing 450065 on 1A28 09.44 Alton to London Waterloo.

455866 leads 455860 working 2U35 12.58 London Waterloo to Hounslow, passing 455863 trailing 455702 on 2G34 12.04 Guildford to London Waterloo at Nine Elms Junction on 9 April 2022. The EMUs are passing Battersea Flyover, which carries the double track Waterloo Curve from Linford Street Junction to Nine Elms Junction, now rarely used following the transfer of Eurostar services from London Waterloo to St Pancras International in November 2007.

455905 leads 455735 working 2O33 13.03 London Waterloo circular via Richmond and Kingston at Nine Elms Junction on 17 February 2018. Nine Elms has an astonishing railway history, with former depots and goods yards on both sides of the lines. On the left, was the North Goods yard and the original passenger terminus (later a goods shed) of the London & South Western Railway until 1848. At that point, the line was extended on the Nine Elms to Waterloo Viaduct to a new terminus at Waterloo. On the right, was the site of the later Nine Elms South Goods yard (in use until 1968), which had replaced a loco works that moved to Eastleigh in 1909, and the mammoth 70A Steam Shed, which remained in use until 1967 – the last year of steam out of Waterloo. This site today is New Covent Garden Market.

In their final week of service, 456013 leads 456021 and 455725 working 2O29 12.03 London Waterloo circular via Richmond and Kingston, passing New Covent Garden Market at Nine Elms Junction on 13 January 2022. Built in 1990–1991 by BREL at York, Holgate works, the 456s have spent their entire life working suburban metro services around London. Their last day in service was on 15 January 2022 and all were moved to Long Marston for a period of storage before being transported to Newport Docks for scrapping.

A busy scene on the approach to London Waterloo as 33207 leads 5Z80 05.53 Southall to Waterloo positioning move with Steam LNER Thompson B1 61306 *Mayflower* on the rear in preparation to work the *Royal Windsor Steam Express*, running adjacent to 442414 and 442406 working 9B88 06.28 Southampton Airport Parkway to London Waterloo with 455738 and 455869 also arriving on a metro service on 20 August 2019.

The Richmond line, Kingston, and Hounslow Loop lines all cross the River Thames and a metro service would do twice in its journey when operating a circular loop from Waterloo. In the new SWR colours, 450052 leads 450058 working 2C20 08.11 Reading to London Waterloo, crossing Richmond Railway Bridge on 22 August 2022. Built originally in 1848, a rebuild was finished in 1908 incorporating much of the original structure but with new steel girders replacing the cast-iron ones. This was one of the first Thames railway crossings. The bridge is situated adjacent to Twickenham Bridge, built in 1933 and carries the A316 over the Thames.

In the new First MTR SWR colours, 444025 leads 444043 working 1W71 15.05 Waterloo to Weymouth away from London Waterloo on 16 July 2022. Waterloo station was the UK's busiest for 17 years prior to 2020, when the Office of Rail and Road (ORR) announced that due to decreasing passenger usage throughout the Covid-19 pandemic, this title had been lost to Stratford with Victoria and London Bridge also seeing higher estimated passenger footfall.

Unlike the 450 Desiros, which can be found on any electrified part of the SWR network, the 45 444 sets are usually confined to express services to Weymouth and Portsmouth. Desiros 444010 and 444026, both in the original stagecoach SWT colours, work 1W06 08.03 Weymouth to Waterloo at Hampton Court Junction in Long Ditton on 27 August 2022. This is where the 'down' Hampton Court line flies over the South West Main Line to Hampton Court. Most of the Class 444s are now in the new SWR blue/grey colours.

SWR 450093 leads 450123 working 1N95 17.54 Waterloo to Farnham as 450052 and 450096 work 1A55 17.55 Waterloo to Alton at Carlisle Lane Junction on 12 August 2022. 455737 and 455720 are in Waterloo South siding following cancellation of a service to Chessington South because of disruption. The first-class seating area on the 450s was moved from an intermediate car to the driving cars and is indicated by the light blue dots (old livery) and the yellow line (new livery) above the window.

458520 leads 458524 working 2K11 07.57 Waterloo circular via Kingston and Richmond pulling away from Queenstown Road, Battersea, under the Chatham lines and passing over the lines towards Stewarts Lane Junction on 9 April 2022. In total, 28 of the 36 Junipers (458501–528) will be sent to the Alstom factory at Widnes for conversion back to four-cars, upgraded for 100mph running and new refurbished interiors suitable for express services on the Portsmouth Direct line.

458523 leads 458535 working 2U33 12.28 Waterloo to Windsor & Eton Riverside between Queenstown Road, Battersea and Clapham Junction on 29 July 2022. The Junipers are a regular feature on the Reading and Windsor lines providing ten-car services. 458531–536 were rebuilt from redundant Gatwick Express eight-car 460 units and all the other 458s were extended from four-cars to five-cars between 2013 and 2016 using a vehicle from a 460 set. The former 460 vehicles can be identified by the continuous ribbon glazing.

After recently returning to service, SWR-liveried 442020 leads 442010 working 9B95 18.48 Waterloo to Poole passing Wimbledon Traincare Depot on 14 June 2019. Sadly, these 5-WES 442s finished at the onset of the Covid-19 pandemic, with SWR later purchasing them from Angel Trains, stripping the interiors at Wolverton Works and subsequently scrapping them at Newport Docks. The Class 442 is officially known as the fastest third-rail train from a time when 442401 and 442403 obtained a world record speed of 109mph between Lichfield tunnels and Shawford, Winchester on a charity rail tour on 14 April 1988.

Shortly after leaving Strawberry Hill, London on the Kingston Loop Line, 707025 and 707017 pass over the Windsor lines, approaching Twickenham Junction with 2K19 09.57 Waterloo circular on 20 April 2021. The Desiro City units would take the 'up' passenger loop in platform 3 at Twickenham, which extends as far as St Margarets Station, where metro trains are often held to allow faster trains from Reading or Windsor to overtake. 707025 would later enter service with Southeastern in Autumn 2021 with 707017 to follow in due course.

It was announced in January 2022 that because of delays with commissioning the new Arterio Class 701 trains into service, SWR would retain 12 Class 707s (707014–024 and 707030) for the foreseeable future to fulfil its timetable commitments. The 707s are often deployed on services on the Shepperton line, which is a branch off the Kingston Loop Line between Strawberry Hill and Teddington. 707013 leads 707002 working 2H26 10.11 Shepperton to Waterloo on the newer 1907-built Kingston Railway Bridge, shortly after departing Hampton Wick on 16 November 2021.

707021 working 2Z19 08.55 Waterloo to Hounslow pulls away from Waterloo on 2 July 2022. This view shows the entrance into the revitalised international platforms that were mothballed following the switching of Eurostar services to St Pancras International in November 2007. To boost capacity at the UK's biggest station, Platform 20 was reopened via a side-entrance on platform 19 in time for the May 2014 timetable change. On 10 December 2018, the concourse and platforms 21 and 22 were reopened on a permanent basis followed by platforms 23 and 24 in May 2019.

701014 working 5Q51 11.25 Staines to Eastleigh test run approaches Chertsey on 23 February 2021. New Aventra units built by Alstom (formerly Bombardier) at Derby Litchurch Lane named Arterio have been on delivery since 2020 but none as of November 2022 have been commissioned for passenger service. In total 60 ten-car 701s (701001–060) and 30 five-car 701s (701501–530) are to replace the Class 455, 456, 458 and 707 suburban fleet. Originally earmarked to all be in traffic by December 2020, severe delays have hampered the introduction of the new fleet. These delays have been attributed to the Covid-19 pandemic, software issues and complaints about the driver's cab interior from train driver's union ASLEF.

Thameslink

Eighty-six dual-voltage Class 319s were built at Holgate Road, York in two batches, 319001–060 in 1987 and 1988 and 319161–186 in 1990 (although many were renumbered during later refurbishments), primarily for use on the new Thameslink network, which was created through the reopening of the abandoned Snow Hill tunnel in 1988. The tunnel had closed to freight traffic in 1969 and passenger traffic in 1916. The reopening enabled the connection of the electrified Brighton and Midland main lines reinstating a Cross-London North-South railway.

The 319s entered service in May 1988 on services between Bedford, Brighton, Wimbledon, Sutton and Sevenoaks, although the Thameslink line in the London core as it is today, featuring a newly constructed tunnel under Ludgate Hill and a new station at City Thameslink (replacing Holborn Viaduct) was not inaugurated until 29 May 1990.

From 2016, new Siemens Class 700s would replace the 319s, although the introduction of stopgap 387s enabled them to be cascaded elsewhere as early as 2015.

Since its inception, Thameslink was an immediate success with trains quickly becoming over-crowded. Starting in 2009, the £7bn Thameslink programme sought to improve accessibility to, from and within London through expanding the Thameslink network, introducing new longer trains and more frequent services. The programme included reconstruction of several London stations, remodelling of track layouts, platform lengthening, building of new infrastructure, upgrading of signalling and power systems, new stabling points and depots and a dedicated fleet of new rolling stock.

GTR a joint venture between Go Ahead Group (65 percent) and Keolis (35 percent) assumed the TSGN franchise from 14 December 2014 and would oversee the introduction of 115 new Class 700s (60 eight-car 700001–060 and 55 12-car 700101–155) financed by Cross London Trains and manufactured by Siemens Mobility. Built at Krefeld between 2014 and 2018, 700108 was the first to enter passenger service on 20 June 2016.

Thameslink 700s first appeared on the Great Northern route working between Peterborough and Kings Cross on 6 November 2017. The GN route was incorporated into the Thameslink network by connecting the ECML to Canal Tunnels, situated between the low-level St Pancras International platforms and Belle Isle Junction just north of Kings Cross, with the first passenger service travelling through in February 2018.

The opening of the tunnels, which had been dormant since being built for HS1 between 2004 and 2006, allowed an increase in service to 24 trains per hour in the London Core from May 2018, with 16 trains using the Bedford line, and eight trains connecting Great Northern stations such as Cambridge and Peterborough with destinations south of London, in Kent, Surrey and Sussex.

From May 2018, the North Kent Line was incorporated into the Thameslink network to increase passenger options for 'cross-London' services with two trains per hour running between Rainham, Kent and Luton. These services replaced Southeastern networkers between Gillingham and London Charing Cross.

There were also plans to expand the TL network to Maidstone East, but this has not yet materialised. The Thameslink programme included a number of major infrastructure projects that feature in the photographs in this chapter including the complete remodelling of London Bridge and the construction of Bermondsey dive-under, allowing Thameslink services to run on their own dedicated set of lines.

The entire fleet of 86 319s worked on the Thameslink route between 1987 and 2017. With just two days in service with Thameslink remaining, 319440 and 319443 work 1G61 07.28 Orpington to Bedford at Elephant and Castle on 25 August 2017. 319435 and 319217 would work the final 319 Thameslink service, which was the 1W46 18.14 Brighton to Bedford on 27 August 2017.

By 18 September 2017, the new Siemens 700s had replaced all Class 319s, 377s and 387s that were used on the Thameslink network. On 7 August 2022, eight-car 700024 works 9O61 17.22 St Albans to Sutton approaching Loughborough Junction.

700111, with NHS keyworkers' branding, powers the 9J09 05.24 Horsham service from Peterborough (the most northerly station on the Thameslink network) at Harringay on 11 June 2022. The train would soon travel through Canal Tunnels, which opened in 2018, just north of St Pancras connecting the Great Northern route to the Thameslink network. Also passing Harringay are LNER Azumas, 801220 with the 1S06 07.30 Kings Cross to Edinburgh service and 800204 with the 5B80 07.21 Bounds Green to King's Cross ECS.

The Sutton Loop Line, also known as the Wimbledon Loop, is a suburban metro route inducted into the Thameslink network from 1995. The Sutton Loop starts at Streatham South Junction with a pair of tracks going south clockwise via Mitcham and a pair of tracks diverging west anti-clockwise via Tooting. There are two services per hour in each direction between Sutton and St Albans using eight-car 700/0s. Under the Thameslink Programme, services were to terminate at Blackfriars to increase capacity in the London Core, but this proposal was rejected by residents. On 28 September 2022, 700045 powers 9O30 10.49 Sutton to St Albans via Tooting at Haydons Road.

The Holborn Viaduct to Herne Hill line on the Thameslink network runs from Herne Hill North Junction to the southern portal of Snow Hill Tunnel, (which then leads into the Snow Hill lines and Widened Lines to Farringdon) following the demolition of the former terminus at Holborn Viaduct. After Blackfriars Junction, south of Blackfriars station, southbound TL services run east via London Bridge either to Rainham, Kent, Brighton, East Grinstead, Littlehampton and Horsham, or continue south towards Herne Hill via Elephant and Castle to Orpington, Sevenoaks and Sutton. 700040 works 9O55 15.52 St Albans to Sutton at Bethwin Road south of Elephant and Castle on 7 August 2022. Interestingly, there are three abandoned stations between Loughborough Junction and Blackfriars Junction – Camberwell, Walworth Road and Borough Road.

In July 2019, ahead of Brighton and Hove Pride, GTR unveiled 700155 emblazoned with rainbow stripes and a heart at each driving motor carriage (DMC) to celebrate the company's LGBT+ communities. Having just passed over the North Kent Line, 700155 *Trainbow* working 9T37 13.19 Bedford to Brighton pulls away from the Bermondsey dive-under near New Cross Gate on 7 August 2022. In the background are Southeastern 375609 and 375920 working 5G65 14.38 Cannon Street to Ramsgate.

Eight-car Thameslink 700057 powers 9P17 08.46 Luton to Rainham away from Greenwich on 9 August 2022. These Desiro City EMUs have Automatic Train Operation (ATO), which is used in the London Core to enable services to run at higher frequencies, giving a tube-style service.

One of the major projects of the Thameslink programme was the rebuilding and remodelling of the lines at London Bridge station. With the new Borough Market viaduct to the west and the Bermondsey dive-under to the east, Southeastern Charing Cross services and Thameslink services now have their own dedicated lines and platforms alleviating congestion. Works on London Bridge started in phases from May 2013, meaning the station could remain open throughout construction. The new station, officially opened on 9 May 2018 by the Duke of Cambridge, now has six terminating platforms (previously nine) and nine through-platforms (previously six), a new concourse unifying the platforms with two-thirds more space and connections, improved accessibility, and more retail and station facilities. With Tower Bridge in the distance, 700133 pulls away from London Bridge with the 9T33 12.19 Bedford to Brighton Thameslink service on 6 August 2018.

When there are engineering works north of the Thames, Thameslink services can often be diverted into London Victoria. Additionally, early morning Thameslink Sunday services regularly start at Victoria instead of Blackfriars, maintaining driver competence. On 3 July 2022, 700023 works 9K90 06.46 Orpington to Victoria over the River Thames at Grosvenor Bridge.

The Bermondsey dive-under is a concrete box structure, which provides grade separation so that the Southeastern Charing Cross trains can dive under the two dedicated Sussex Thameslink tracks without impeding each other. This new structure ultimately 'untangled the tracks' on the eastern approach to London Bridge, improving the flow of trains and increasing capacity. Southeastern 376030 and 376005 emerge from the dive-under with the 2F48 15.35 Sevenoaks to Charing Cross service as Thameslink 700127 flies over with the 9S40 15.12 Brighton to Cambridge service on 14 September 2022. Work on the dive-under started in spring 2012 and was completed in late 2016 with the first passenger train running through on 3 January 2017.

Other books you might like:

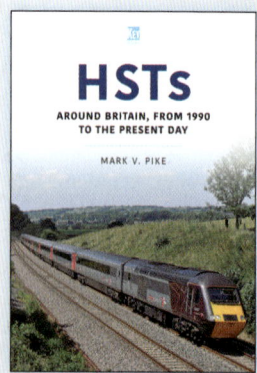

Britain's Railway Series
Vol. 33

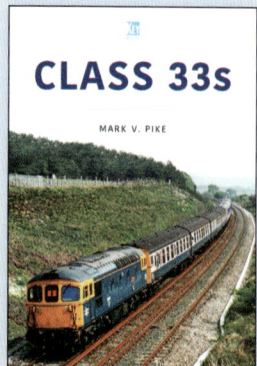

Britain's Railway Series
Vol. 40

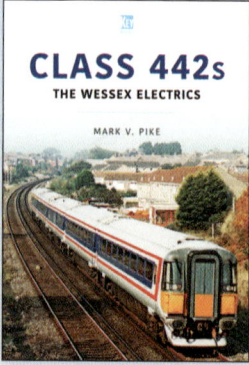

Britain's Railway Series
Vol. 27

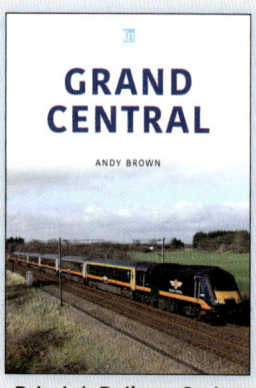

Britain's Railway Series
Vol. 31

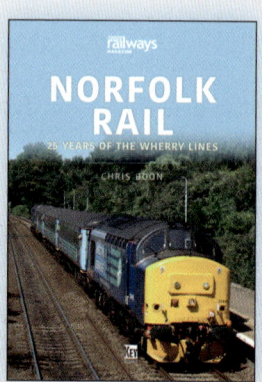

Britain's Railway Series
Vol. 20

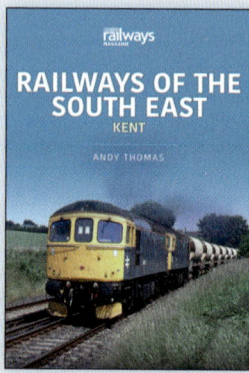

Britain's Railway Series
Vol. 15

For our full range of titles please visit:
shop.keypublishing.com/books

VIP Book Club

Sign up today and receive
TWO FREE E-BOOKS

Be the first to find out about our forthcoming book releases and receive exclusive offers.

Register now at **keypublishing.com/vip-book-club**

Our VIP Book Club is a 100% spam-free zone, and we will never share your email with anyone else. You can read our full privacy policy at: privacy.keypublishing.com